Conversations
with Birds

"The strength of Alan Powers's own connection to the world of the natural is never in doubt, and the wild birds he sees and hears inspire him not only as a writer but also as a critic and a musician. His discussions of poetic and dramatic texts by Keats, Shakespeare, and especially Dickinson are fresh and enlightening. What the essays, anecdotes, musical citations, and literary musings all share is the conviction that 'birdtalk'—the habit of conscious observation and reflection on the connection between the human and natural worlds—can save us from the frenzies of life in the age of technology. There is no sentiment, no pathos in the lesson. Powers does not force it on us but rather offers it here for us to take to heart if we choose, but I cannot imagine the reader of this heterodox and delightful book who turns from it unmoved."

RICK WRIGHT, AUTHOR OF *PETERSON REFERENCE GUIDE TO SPARROWS OF NORTH AMERICA*

"*Conversations with Birds* is a wonderful book filled with great insight into the nature of birdsong and our own birdlike inclinations. Many bird books seem too dry and scientific, but the musician and poet come out strong in Alan Powers and speak vividly and in a way that appeals to the artistic type. It is obvious that Powers is a poet and a lover of poetry. I really enjoyed the chapter on literary birds. I read this book like a devotional—mostly reading it in the morning with coffee while watching and listening to the birds at the feeders in my backyard. Thank you for a joyful and inspiring book! I'm going to promote this book among all the other bird-loving rockers and poets I know."

JERRY OLIVER, SINGER-SONGWRITER, INDIE ROCKER, AND CREATOR OF THE BIRDWATCHER EXPERIMENT

Conversations *with* Birds

The Metaphysics of Bird and Human Communication

A Sacred Planet Book

Alan Powers

Bear & Company
Rochester, Vermont

Bear & Company
One Park Street
Rochester, Vermont 05767
www.BearandCompanyBooks.com

Bear & Company is a division of Inner Traditions International

Sacred Planet Books are curated by Richard Grossinger, Inner Traditions editorial board member and cofounder and former publisher of North Atlantic Books. The Sacred Planet collection, published under the umbrella of the Inner Traditions family of imprints, includes works on the themes of consciousness, cosmology, alternative medicine, dreams, climate, permaculture, alchemy, shamanic studies, oracles, astrology, crystals, hyperobjects, locutions and subtle bodies.

Cataloging-in-Publication Data for this title is available from the Library of Congress

ISBN 978-1-59143-451-1 (print)
ISBN 978-1-59143-452-8 (ebook)

Printed and bound in the United States by Versa Press, Inc.

10 9 8 7 6 5 4 3 2 1

Text design and layout by Priscilla Harris Baker
This book was typeset in Garamond Premier Pro with Futura, Masqualero and Snell Roundhand used as display typefaces

To send correspondence to the author of this book, mail a first-class letter to the author c/o Inner Traditions • Bear & Company, One Park Street, Rochester, VT 05767, and we will forward the communication.

For Sherley and Leonard Unger

For Susan Mohl Powers

We two alone will sing like birds i' th' cage.
—SHAKESPEARE,
KING LEAR, ACT 5, SCENE 3

Some reputable scientists, even today, are not wholly satisfied with the notion that the song of birds is strictly and solely a territorial claim. It's an important point. We've been on earth all these years and we still don't know for certain why birds sing. We need someone to unlock the code to this foreign language and give us a key; we need a new Rosetta Stone.

—ANNIE DILLARD,
PILGRIM AT TINKER CREEK

Contents

Foreword

Richard C. Wheeler

*L*ike Alan Powers, I am a bird person. Both our lives have not only been touched by birds they have also overlapped with birds. Al whistles to birds and communicates with them, and I have rehabilitated birds for about sixty of my more than seventy years. If a bird needs a warm place and sustenance while recovering from a minor injury or ailment that doesn't call for a vet, I am its harbor of refuge.

I have often walked with Alan when he talks to birds, on the roads and shore in Horseneck Bay not far from his home in Massachusetts. He hears a call, listens, then whistles, cocks his head, listens again. Soon there is what might be called a conversation, as Al listens intently for the bird to answer and judiciously enters his own bird statement.

What is he saying? What are they saying? He hazards some guesses, in this chronicle of his life as a whistling bird communicant.

In the equine world there are the horse whisperers. In the avian world we have Alan Powers, the bird whistler. When Alan whistles "like" an Oriole, he is an Oriole—but he is also aware of and questioning his own place in their world and whether his presence is intrusive. He has thought about the sound not just as sound (and he is a musician, who plays the notes on his piano and labors to transcribe the riff, comparing it to classical themes he knows) but as communication. "What does the bird mean when it says that?" he asks and sets out to find the answer.

Most ornithologists or binoculared birders love the songs of birds. There are many books and CDs on birdsongs. Who has not felt a thrill at the first call in April of a bird not heard since October? But few of us have the skill as whistlers to answer back, to reproduce a birdcall almost perfectly, much less to sort out who's talking to who.

Powers has a dry humor when he writes about birds. He takes the perfection of his bird whistling seriously, but humor is his métier. His birdcalls inspire allusions and references from Shakespeare, Emily Dickinson, and Robert Frost, from English and American literature, from the Italian Renaissance and popular culture, and from bird lore and natural history. He listens to birds and replies, in the city as well as the country, in England, Italy, Europe, and around home in New England.

*BirdTalk** is a close-up and exciting visit with a man of enormous erudition and humor who finds the natural world splendid. Chapter drawings of each bird are contributed by Susan Mohl Powers, which will help you identify the bird whose call Powers has been describing. You will delight in Alan's thoughts and Sue's drawings and then want to go out and try to whistle back to birds yourself.

When I take care of birds and nurse them back to health, I have been their nurse, yes, but also their jailkeeper. They have been stuck with me—the bond between us based on food and

---

Conversations with Birds was originally published as *BirdTalk* in 2003 by North Atlantic.

shelter. Alan knows how to knock, ever so gently, on their doors, and sometimes gets invited in. For that, I kind of envy him.

RICHARD C. WHEELER

MARION, MASSACHUSETTS

RICHARD C. WHEELER spent his youth hunting and fishing in the marshes of Marshfield, Massachusetts. He earned a graduate degree from Harvard University and devoted his life to education and conservation. He was a former Navy SEAL, he was curator of the USS *Constitution* Museum, has been the headmaster of a number of New England secondary schools and later served as director of the USS *Constitution* Museum and Cape Cod Museum of Natural History. In 1992, at age sixty-one, Wheeler kayaked over a thousand miles from Newfoundland to Buzzards Bay, retracing the migratory path of the Great Auk, an extinct bird. He made a much-praised film, *The Haunted Cry of a Long-Gone Bird,* using the extinction of the Great Auk as a metaphor for the collapse of Newfoundland fisheries. *Time* magazine named him a "Hero of the Planet" in 1998. He passed away in Boston in 2019.

A Preface to the New Edition

Talking to Birds and Aliens

*A*ll living beings communicate with their own species and react to other species, often with fear—as in the common gargle-rattles birds make when detecting a threat. Reversing this species-centric tendency, I speak with other species and fear my own. Communing cross-species, I reach out to what Giordano Bruno, a Sixteenth-Century Italian Dominican friar and philospher, called in Latin the *universus,* the "turned into one."

My book *BirdTalk* gained worldwide fans, among them Hollis Taylor, an Australian violinst, composer, and zoo-musicologist in 2003. I am not alone in my fascination with birdsong and my efforts to imitate and notate their music. Taylor has composed music based on Pied Butcherbird intervals. Both A. J. Mithra, from Chennai, India, another zoomusicologist, and I have written tunes based on the Carolina Wren.

My musical instruction from birds and my musical development have yielded fruitful effects on myself, including health of spirit. A mind at peace with itself protects the body's health, according to Thai healer Tulku Thundup (1998, 12).

For a mind at war with itself, as my critical mind can be, tune-talk restores peace.

Humans consider bird speech as music because our own speech lacks song, except from great poets. When birds speak their worries—about weather, about location, about food—they contort their vocalization. Still, even their anguish and anxiety sound like music to us.

We may respond and engage, though it takes practice; for instance, talking to small English Robins. To prepare before crossing the Atlantic, I whistle as high as I can, for a month. One of my proudest conversations took place in Kensington Garden, London. A sustained, high conversation with one English Robin, ten feet away. They're small, but feisty.

Crows, like all Blackbirds (Starlings, Ravens), boast complex communications. English Blackbirds talk lower; they slur in whistling range, in diatonic (piano) intervals. Visiting Weymouth, Dorset, I held a five-minute conversation with such a Blackbird on a nearby stone house roof. I even notated seventy measures, some repeated, on musical staffs.

Morris Swadesh considers the origins and diversification of languages, including Maine dialect "baan" for barn (2006, 15). My having admired the irony of Maine talkers like my grandfather, I say my failure to roll the Italian *r* is actually a Maine accent in Italian. Meanings diverge geographically with human and bird talk.

Birds arise in love songs throughout the world, as in the Neapolitan song *"Nu passierello sperzo,"* the sparrow chased away by the birth of a child or by lovers in the woods (Paliotti 1992, 92).

A. J. Mithra and I both wrote tunes based on the Carolina Wren, often a triplet talker. Of tripleness, Leonard Bernstein says, "It's true that much music is triple. The three-concept is almost as fundamental as the two-concept. But three is not grounded in our biological nature. It is not physical in function. The heart just doesn't beat in 3/4 time, Viennese propaganda to the contrary" (1966, 95). Adding a triplet to a duplet, we have a five-note series: maybe not human, but birds with two wings, two feet, and a tail? Maybe triplets are physical with birds. And quintuplets.

The most beautiful birdtalk, the pentatonic song of the Wood Thrush's, like the five black keys on a piano, often ends in a rasp. I say, all birds want to be drummers, like insects or octopuses. Many like to imitate other sounds, garbage cans or children playing. One Macaw imitated my laugh, exactly. Pied Butcherbirds sound like Aaron Copland with tritones and octave-plus intervals. When a Blue Jay imitates a Red-tailed Hawk, he terrifies his victims into flying where he can see them.

Of course, the cosmos fills with sounds, as does the universe, planets and beyond the heliopause, as in Voyager 1 plasma waves. I don't plan to communicate beyond our humble planet, though I listen to the wind on Mars from rover *Perseverance*. In college I wrote fifty pages on sound and meaning: reading Renaissance poets Wyatt, Spenser, Donne, and Milton. For each poet, sound entwined in a similar way in their best and worst passages: for example, Milton's tendency to move from rising iambs to falling, didactic trochees or even dactyls: "and with the setting sun / Dropt from the Zenith like a falling star." Commands and moral demands employ a falling rhythm, "LOCK him UP" or trochees, even, MAKE SURE YOU. Historically, the Nineteenth Century often used falling meters, dactyls like "This is the forest primeval," and that century preferred didactic verse, issued from above, the rulers or higher.

Does this work in birdtalk, where falling vocables like the Titmouse's early glissando, or the Cardinal's, or the Red-tailed Hawk often imitated by Blue Jays?

Time and the uni-verse will tell, in my book.

Mourning Dove

Basic Bird
A Prologue

Mastery of a half-dozen sounds and intervals enables the enthusiast to talk with birds. The large fish eaters, Osprey and Bald Eagles, make a high sound like sucking in through your teeth, lips somewhat pursed. Wetting your whistle can help. Be careful how you use this powerful though faint, high sound. The hallmark of a predator, it is not universally welcomed by smaller birds.

Still with your lips pursed, form the sound of a *T* by tonguing against palate or teeth. This sounds almost like *tsk, tsk* but is more plosive or explosive. It's the sound most birds make when warning—whether other birds or their own offspring.

Now we advance to whistling. Start with the minor third, usually descending, like the first two notes of our national anthem, "O-oh." Stop before the "say," which turns the whole into a major triad. Just the first two notes, "O-oh." Now whistle it. This is perhaps the commonest avian interval, shared by many varieties from the Titmouse to the Oriole to the White-throated Sparrow (or Peabody Bird), which more often sings an ascending minor third. The White-throat inhabits pine forests like those in Maine and Massachusetts.

Next to the minor third, both on the musical scale and on

the scale of frequent birdsong, is the fourth or fifth. Descending, it's a common, strident call of the Mockingbird and Cardinal. Ascending, it's usual in Robin songs, especially between the first and second part of its burbling chatter. Most Robins sing fourths, but in certain areas (like Providence, Rhode Island), I've heard individual birds sing perfect fifths. The fifth is the interval in the national anthem between the first "O" and the "say." Omit the second note of the "O[oh]." Now whistle "O[-] say." That's a descending fifth. If you slur or glissando between the two notes, you have made a fair approximation of a Cardinal call—that is, if you've whistled high enough, in the Cardinal range, around F above the C above middle C on a piano. Even the familiar, bluesy, regretful song of the Mourning Dove flutes an ascending fourth, and then falls back to the half tone higher as a grace note. Of the Eastern American birds, the tonal register of the Mourning Dove most resembles Native American alto flutes of the Southwest written an octave higher.

Oscine passerines make many other sounds. (This taxonomic term covers most songbirds, *oscine* because they were once used for divination, *passerine* for "perching.") Some of these you can find ways to imitate. Many I find entirely outside my capacity. Gulls I can imitate by swallowing or gulping. One Blue Jay call combines a similar gulpy effect with a flutelike sound. It's a common call note, especially in the fall, but I find it beyond my skill so far.

A word about rhythm. Some birds seem to be drummers: Pileated Woodpeckers, Hairy Woodpeckers, and various game birds like the one we heard every evening on a college hike in March. We called it the Blue Ridge Thumper (it turned out to be a Ruffed Grouse). Like the Song Sparrow's, its rhythm could be represented only by complex math, but it was fairly easy to imitate when we heard it. For the rest, musical notation gives a fair analogy to many standard bird rhythms: say, the dotted quarter note followed by an eighth note, roughly *dah-di*. Many birdsongs sound syncopated, as if their ac-CENT is on the wrong syl-LA-ble—especially at the beginning of a birdcall. I have represented this effect in musical notation by starting many birdsongs with an eighth-note rest.

Use caution when responding to birds. They do not possess the same instincts or impulses we inherit from our apelike forebears. For instance, keep your distance—say, thirty feet—when imitating a birdsong. For most birds, twenty or thirty feet is roughly equivalent to a yard among humans. Closer than that, and you are up close and personal, within the bird's urge-to-flee space. Don't talk to a nearby bird unless you can disguise yourself as a twenty-gram ball of feathers.

Also, be careful about rising versus falling glissandos. Under certain circumstances, the one may be an enticement, the other a shooing-away. One mid-March morning I heard a Titmouse making its usual challenge, a quick descending minor third. I responded with its other call, an ascending minor-third glissando one half tone lower. It returned my call and flew away a bit, maybe ten feet. With each repetition it kept retreating, I think in a "come hither" manner. When I tried its original descending challenge call, it did not repeat it. Both calls asked me to appear and identify myself, but the descending suggested a territorial dispute.

I assert here that almost never are two birds' calls exactly the same, so this book concerns ear training, as well. They may sound alike at first, but the more developed your ear, the more individuality you will find.

Oriole

The Year
of the Oriole

I am no saint, but I do talk to birds. They respond, most of them, depending on time of day and time of year. I am not sure I should tell you what they tell me. It is privileged communication, "For ears only," as government security might put it. It is a secret, an open secret.

One thing I am at liberty to reveal. They want in to sports competition. Birds feel that we lesser great apes have for too long bathed in our own simian glory. We have celebrated our own species, one limited, after all, to treading Earth. I've met Hummingbirds who would win all the Olympic dash events—100 meter? 200 meter? No problem. As for high hurdles, the Hummers laugh at them. Not very high. And no hurdle at all. Birds are not alone in their desire for cross-species competition. During high school summers, I practiced running dashes on my grandparents' dirt road, a 7 percent grade near Norway, Maine, and then in autumn out back on a suburban soccer field. The dirt road held just enough incline so that when I returned from vacation to run on the level, it was easy, like running downhill. But at home on the level I picked up some real competition, my neighbor's French poodle, Tart. That little curly haired fancy prancer ran fifty yards in under five seconds. 'Course, Tart never ran the full one hundred yards straight; he always ran toward the hash marks, then to the opposite sidelines, and then cut back. A great broken-field runner, Tart's football career was hampered because he could hold only a kid's tiny football. For Olympic competition, the Interspecies Olympic Committee will have to certify dogtrack equipment—a fake rabbit or real ball for canine entrants, a bucket of money (or perhaps a news camera) for sapiens.

And sugar-water for the Hummingbird entrants, since we are talking here of birds. As I write, it's five o'clock on a July morning. The birds have just begun. Robins—it has been rainy, so they sing more—and distant chatterers: Sparrows, Titmice, Finches and Warblers, Catbirds and Mockingbirds, even Doves and Crows. Now join in the Carolina Wrens. They pipe variations on the "Mexican Hat Dance," the first part (or in classical harmony, the second inversion of the subdominant, fourth, major triad). Over and over. Sometimes they start on the last note of the "Mexican Hat Dance" triad. One day in late July I hear them do simply the first three notes of a major scale, rapid triplets. Over and over. At this time of year, the variations are freer. One sounds almost like a Cardinal, that glissando of a bird. There's the Cardinal, now. Today it's piping whiplike rising glissandos in rapid succession. The top note of the call lowers a quarter tone with each gliss. Oh, there's the rising staccato of a Robin on takeoff. Later in the year, "across the pond" in England I hear a Blackbird or continental "Merlo" make a similar flight song. It's in the same genus, *Turdus,* as the American Robin.

As I said, I'm no saint, and now I must add, nor even a Cardinal. Then again, many of the saints weren't saintly. Not by midwestern standards. Ever read Augustine's *Confessions?*

Nowadays he might be jailed for failure to pay child support or have his wages—as professor of rhetoric, first in Rome, then in Milan—attached. Think of it, sending the mother of your child back to Africa because you're about to marry, and then taking up with a new mistress because the marriage is a while off. Augustine was not as faithful as a Goose, although some passerines would sympathize.

Now they have stopped, the birds. Why? The sun has not risen, though it's daylight now. A month ago they would still be singing this long after dawn. What are they up to? In spring they help the sun rise; they levitate it with song. It's their paean of praise. It affects the heavens. But now they have stopped.

Assisi Species

They stopped. That's what they did when St. Francis preached, according to the early accounts. Brother Leo, in the *Speculum* or *Mirror of Perfection,* reports two bird encounters in an otherwise busy, charitable, if anti-intellectual life. (St. Francis forbade books, even a breviary to "Friars of mine who are seduced by a desire for learning" [Habig 1991, 1198].) Near Bevagna, on his way from home base at Porziuncola to Spoleto, Giovanni di Bernadone saw a field below the crenelated city walls, a meadow filled with birds—Crows, Rooks, Wood Pigeons, and Orange-throated Bullfinches. As he approached, Giovanni/Francis did not spook them. They stayed. In Julien Green's account, not even a Magpie budged (Green 1983). Even when Francis started preaching to them, they did not leave. Of course they didn't understand Italian, though it is closer to birdsong than English. He congratulated them on their dress and spoke of how they should praise God for their existence.

One had thought *they* could tell Francis how to do that. At any rate, the birds stayed until the saint dismissed them, so we are told. It is not too far-fetched to imagine his dismissal a sweeping gesture with his cloaksleeves.

What does this story suggest—besides the writer's anxiety about holding congregants' attention, a sensible anxiety because St. Francis insisted that his followers not study the art and technique of preaching? First, it tells something about time of year and time of day. Chronobiology. Birds flock together at various times, for various mysterious and not-so-mysterious reasons, such as migration. There is even a German word for premigration nocturnal flitting, *Zugunruhe.* If you were about to leave, like the Arctic Tern, on a ten-thousand-mile flight under your own steam, you might raid the icebox or flit about, too. Second, the account of St. Francis at Bevagna tells us about his manner and his dress. I have known one teacher comparable to St. Francis in this regard. He spent one summer tenting in Vermont's Bread Loaf Wilderness, his campsite part of the extensive realty claimed by a female moose. His Complete Shakespeare—almost pocket-sized, in eight-point type—was wrapped in a plastic bag. Tall and darkly dressed, on the trail he blended in with the trees. His motions were circumspect. The moose had checked him out, as St. Francis's flocks had the saint. He was "OK."

I visited Assisi one January morning. Huge industrial cranes rose over the town, the aftermath of the earthquake a few years back. Many farms now double as hotels featuring agrotourism. From a bridge over the Tiber—here, a forty-foot-wide stream—I could see Dabbler Ducks and a brilliant black-and-white Goose. Stopping in the lifting morning haze, I saw small birds gleaning in the fields and flitting to small fruit trees. I talked with the great-great-great-great-great-great-great-great-great-great-great (this could go on for a page, since bird generation is so rapid) grandchicks of St. Francis's birds. And indeed, they did bespeak gratitude for the sunlight, the mild January weather, each other. Whether they learned it from the saint, I cannot say. If they did, one suspects they did not have the message quite right after all those years of oral transmission. (Recall the childhood game of "telephone.") All I said to them was a forceful, in-sucking *cheep cheep* like theirs. They seemed to be some form of Sparrow.

Zugunruhe. Back to my July birds' 5 a.m. silence. Do the many migrants among them, the Cardinals (in the Northeast), the Orioles, the Robins, the Hummingbirds, wonder at the sluggard sun after their winters in Georgia or Florida or Venezuela? The most stunning silence from birdsong I witnessed, of course, during a solar eclipse. I cannot now recall whether it was a silence made loud by increased birdsong as the sky darkened. That would make sense, so it may not be true.

As for St. Francis, shall we say his birds treated him like an eclipse? Possibly, although probably more like a denizen, like the camping scholar on Bread Loaf Wilderness. With his simple clothing and reserved movements, and no hunting cap, St. Francis's gestalt or aspect was natural. Funny the difference a hat can make. On my morning walk I often stop to give some clover to two donkeys. I don't talk to them because a donkey's conversation is a spectacular mating eruption, a volcanic, sawing in-exhalation or *hee-haw* that denies meaning. His words are naught but the braying of asses, as the Bible says. So one morning I wore a new wide-brimmed sunhat. The donkeys would not approach. Who was this galactic visitor with the Saturn-ringed head they'd seen in their worst dreams?

Careful What'cha Say

Take your palm and tap four beats steadily on something in front of you—a table, a book, a PalmPilot. This morning, early, I heard a call note in that rhythm, three notes and a half-beat delay on the fourth, a descending minor third. *Deep, deep, deepuh dee.* (The *uh* is a musical rest, silent.) I don't know which bird it is, and I'm in bed at 5 a.m., unwilling to find out. Though I learn some dozen new songs a year—all for the same bird—there are thousands of songs I don't yet know. These occur in my own region, to say nothing about a province away. Like all language learning, I immerse myself in the unfamiliar until over the months and years it becomes familiar.

I have declared this the year of the Oriole. (Sounds like the desperate optimism of a baseball fan, but no.) Orioles make a plaintive call note, a falling half-tone glissando. I believe they also make it on the wing, like Robins' staccato patter taking off. A few years back I pulled a St. Francis and tried to teach the birds. Recall that St. Francis of Assisi's peculiar gift was not listening to birds but "preaching"—the word used in the early accounts—to them. Professionals easily credit this story. Anyone who preaches or even teaches can believe that the birds make as attentive listeners as most parishioners or college sophomores.

This Oriole year I gave up trying to teach them a famous symphonic riff. (Stay tuned.) I just listen, and if they are far enough away, I echo them. You must be careful if they're close. An echo at close distance chases them off. Probably half of all birdtalk, like people on cell phones, is locative. "I'm here. Where are you?" Some birdcalls function as simple greetings, others as choral matins, telephones, road signs and roadways, flight paths, roadside route numbers, vegetable-stand signs. At the same time they can be stylish vehicles—elegant dress or ceremonial robes—performed to impress. Season dictates much. Spring call notes roughly translate, "Get out of my tree!" They just say it with much more style. "Yo! Flee the tree, see?" But on a cool fall morning, the Cardinal's quick-rising, whiplike call note finds a distant answer. Is it his broodmate? His colleague? His son? Will one of them migrate, or did they both inherit nonmigrating genes?

Back to the summer Oriole, who hits a plaintive call note. He follows it with a cheery song, the rhythm of *"Gaudeamus igitur"* (the college Latin drinking song). The melody is a major triad starting on the second note. Inexplicably, the great Nineteenth-Century American "birdear"—pun intended— Simeon Cheney, in his enthusiasm, demurs about his song, calling the Oriole "hardly a songster."

> He is the most beautiful of our spring visitors, has a rich and powerful voice, the rarest skill in nest-building, and is among the happiest, most jubilant birds. Hardly a songster, the oriole is rather a tuneful caller, a musical shouter. (Cheney 1892, 72)

Cheney does not appear to have read Thoreau's journal encounter with the "ones that Midas touched," nor to have heard the Warbler-like chatter I describe a page farther on.

Symphonic Bird

I don't know how he did it, Giovanni di Bernadone (St. Francis, to you). He seems to have engaged them somehow, spoken *to* them. I tried this once—with Orioles, too. I noticed one who made three short notes and then two glisses a minor third lower. To hint at what I tried to teach him, the first three notes sounded like the first notes of Beethoven's most famous theme, starting his Fifth Symphony. The only hitch was the fourth note glissandoed down a half step. And the Oriole's fifth note repeated that gliss exactly. I figured I could use the bird's tonal flexibility and whistle the fifth note a half step plus a quarter tone *higher.* Bigawd, the bird did it! I had "preached"—what, music practicum, I suppose—to a bird. His four notes sounded almost exactly like the first four of Beethoven, except his glissando on the fourth, down from the minor third to the major. And, truth to tell, the fifth note was a quarter tone lower than Beethoven, but "close enough for union work," as musicians say. I repeated my "prompt," and he repeated his sluicy Beethoven. About three times. But having succeeded with the one note, of course I advanced to the descending fifth with which Ludwig concludes his phrase. And I heard . . . I heard . . . nothing. I tried it again. Nothing again. That last blew his mind, or it blew my cover as an ersatz Oriole. At any rate, the Oriole could not conceive or perform such a note in such a sequence. It was uncanonical, outré, a gaffe, a breach, a solecism if not a faux pas. I may have sounded to him like an overbearing megalomaniac crazy to close the conversation, for after all, who was he talking to? Beethoven, not me. That eclipse of a composer.

I should add that the Oriole's instinct was very like the German composer's. That descending fifth interval comes much later in the symphony. Beethoven, like the Oriole, makes a long series of *dih-dih-dih dahs* in different keys before he gets to the descending fifth.

Perhaps less than one-fifth of all birdsong is imitable by humans. Although I write about that one-fifth, I do not forget the vast ranges of inimitable birdsong: almost all songs by Blackbirds, Warblers, Finches, scavenging shorebirds, Hawks and

Kestrels, even Wrens. And hundreds of varieties besides. Even among the imitable birds, there are often inaccessible languages. Partly this is a matter of higher frequency than most humans—at least, than I—can whistle. For instance, Orioles have a whole other language they shift into after they have given their "dial tones," or series of them. Their second language sounds high pitched, like a Warbler or Swallow or Red-winged Blackbird, though it lacks the overtones of the latter. This second, Oriole-to-Oriole "specielect" must be what Thoreau witnesses in his *Journals,* where he calls these birds "Golden Robins" (or because of their nests, "Hangbirds"). Thoreau records, *"May 8, 1852* Two golden robins; they chatter like blackbirds; the fire bursts forth on their backs when they lift their wings" (Allen 1993, 287).

It's difficult to know what birds are saying, though something can be gleaned by context. Admittedly, an observer's preoccupation colors interpretation. My friend, a father of nine, thought that the summer birds near my house were asking me to feed them; I agreed that they were discussing food indirectly, by staking their territorial claims. Emily Dickinson in one poem goes so far as to "hear" moral values from a Robin. The Mourning Dove gets its name from the mournful affect of its song for us humans, whereas it may be singing of pure joy, "like to the lark at break of day arising from sullen earth," as Shakespeare says in his Sonnet 38.

The first time I really held an extended back-and-forth bird conversation—echoic, granted—was with a very aggressive Titmouse. Probably because of the demands of specialized feeding and population density, Titmice are among the most aggressive local birds, so you can appear to "call" them simply by reiterating their call note. It's a descending minor third, repeated usually three times. Return this call insistently and a Titmouse is sure to appear on the tree above you. When he does, you will know that what you said—essentially, "Get out of my tree"—is exactly what *he* was saying.

Birdtalk tells us much about our own human speech. Our greetings, for instance, share something of birds' call notes. Think of a host's greeting at a toney party, say, in F. Scott Fitzgerald's novels of the Roaring Twenties. "How d'yi do?" is a greeting, yes, a welcome. But it's also a territorial warning. Whose turf are you on? The host's. What *is* the "territory" of a human call note? It's a vast social network. Where are you in this network? If one considers the implications, an introduction, in that old, static society, was a crisis. Best get out one's clearest whistle and pipe a repeated minor third, like the Titmouse.

As I write this on a July morning at the beach, I hear the *plop* of what sounds like a fishing lure some three rods "down-sun." (We have the perfectly good downwind, but for birds, we also need solar-oriented adverbs.) Turning, I see a Tern rising. It fishes on calm waters by throwing itself like a lure. Can't see whether it hooked or taloned anything. Two Swallows skim the irregular wash or scree, small boulders that stretch a mile here. Three or four rods up-sun, out past a pair of sapiens, a Cormorant surfaces, looking huge in the flat water, like a sea serpent on the borders of Renaissance ocean maps. Cormorants are the "Buzzards" for which this bay is named, Buzzards Bay. In flat water, the Swallows skim the water, too, for their favorite bugs. Mmm.

Levitating the Sun

It's 5 a.m. again, a couple of days later. I am home, listening to the first few birdsongs of the day. Out front quite close—maybe in the shrubbery—a Cardinal starts its rising, whiplike whistle, like the first part of the old-fashioned wolf whistle. (You still occasionally hear *Avis sapiens* perform this in situ at a construction site.) Anyway, the first, rising part of the wolf whistle—the *wheet* in the *wheet-whew!*—approximates this Cardinal call note. This morning, I delight to hear him build first isolated whip-like *wheets*. Then two together. Then two together followed by a falling, Oriole-like plosive. Then two whiplike *wheets* followed by two plosives, or what I call "punctuation." If we could whistle punctuation, these would be exclamation points. Then two whiplike wheets and a succession of exclamation points—four. Next time, two. Then four—plus three. Like a jazz musician, he's building riffs. And he's also warming up like a jazz musician, or an after-dinner speaker. Meanwhile, the sky is ablaze with pink. He stops. He's done his job.

During this pause I should talk about gender and birds—the birds and the bees, if you will. I have used the masculine pronoun, and with a Cardinal it's easy enough to tell the bright-red, higher-crested male when you can see him. The whole point about birdsong is, however, that most of the time you can't see them. That's the fun, and the challenge. There's a brief period in spring before the trees have leafed fully, when, if you work at it, you can chase down many singing birds and see them. During this brief, leafless period I have found the specific trees or fields along my road where the Song Sparrow sings, and the White-throated Sparrow, the Catbird, various Warblers, and the Wood Thrush—though since I first identified a Thrush, in my teens, I have almost never seen one singing. They do it in the tops of tall trees at dusk or dawn or whenever it's hard to see them. Very like Robins in the rain.

But back to gender. It's a grammatical convention, and a choice. For several species with which I am familiar, a preponderance of male avians sing. Or chatter. Call it bravado. Even threats. For birds do live in a dangerous predatory world. Even the smallest—especially the smallest—participate. Their "participation," to excuse my euphemism, is often as breakfast-egg producer or as food. Birds live in a hostile world. Anyway, I will use the male pronoun except where I have observed female birds singing.

A little after 5 a.m. again, and there's more chatter today—Warblers, it sounds like, busy and high, much out of range of imitation except by sucking in through one's teeth, crude and imprecise, but at higher frequencies than I can whistle in the usual manner. Maybe a Catbird, doubtless a Mockingbird, no Robin very close anyway. Their burbling, emphatic call—"hurried, few, express reports" as the poet Dickinson says—often dominates. As often they alone sing, or "play," because singing implies an expressive content that suggests joy. Even as I note their absence, I hear one, on the other side of the house, and some thirty minutes later. (No, it did not take thirty minutes to write this paragraph. I also put on the tea kettle and poured a cup.)

That in-sucking sound is like a sharp knife, to be avoided until you know its correct use. When I suck in one clear high note, it's as close as I can come to the Osprey and Hawk (and Eagle) call. Avoid it unless you want to send flocking birds—Starlings, Seagulls—scattering in caution or terror. If I in-suck a moist overtone, the sound approximates the most common of avian warnings made by all the passerines I am familiar with. One mild mid-February day I found the Titmouse on one side of my road following my mimicry along with its characteristic descending minor third. On the other side of the road, a Titmouse followed in the bushes without using its familiar call note. Instead, as it retreated a couple of bushes, it scolded in the plosive variant of that in-sucking warning. Near it was another Titmouse. Had it already paired on February 16? Apparently. Maybe the old British folklore that birds pair by St. Valentine's Day bears a flake of truth.

Now it's late morning, 11 a.m., and late July. Very humid air, 90 degrees Fahrenheit, while the ponds and creeks are nearly dry. I pass a Snowy Egret standing on the only rock in the local millpond, the only rock still surrounded by water. Maybe a foot deep. Two Egrets, three Canada Geese, and a few Mallards grace the pond's edge. No Glossy Ibis with its long curved beak this year, though one was here the past few years. Hope he or she is OK. Talking to the birds makes one see their lives and their struggles. Each one over a year old does have an idiolect, its own way of communicating—each Robin, each Cardinal. Near the Ides of March I hear at an ATM a unique Cardinal in a bare tree. He whistles a rising triplet with a sixteenth rest between the first two slurred notes, C and D above middle C. Then he accents the fifth, the G. Of the hundred Cardinals I have heard, no other sang exactly that song. The ancient Romans might find it auspicious.

I just heard a sound, a short glissando from fairly high in my whistling range, too high for an Oriole. It could be the second part of a Robin's burbling call notes. But no, he then follows it with five, then six notes. When I imitate these, he makes the distinct, longer (but still truncated) Cardinal gliss. Twice. Then he follows with the quintuplets, then sextuplets, descending a quarter, then another quarter tone. Is this an idiolect, specific to this particular bird, or specific to midsummer, late July? Or to this time of day, late morning, and this hot weather? I have conducted no double-blind study to eliminate variables, so I must leave it a question.

Not Just Babel

One thing is clear. Many birds make some similar sounds mixed in their call notes. Descending minor thirds are whistled by Titmice, Orioles, Robins, Chickadees, White-throated Sparrows (though usually those are ascending), Catbirds, and Mockingbirds, to name a few. But each has an entirely different timbre or sound quality. The Catbird's timbre is "thinner" than his cousin the Mockingbird's. Both the Titmouse and the Carolina Wren repeat insistently, obsessively. But the Wren varies his call notes more, and they are triplets, often syncopated. Although Robins are repetitive in their way, their notes are impure and sidling, burbling, so they don't drive you mad in their iteration.

Contrast Titmice and Carolina Wrens, which might be used like water torture. (See Hartshorne 1956 on the monotony threshold in songbirds.) This only makes them easier to learn. Like calisthenics and exercise programs, or like piano scales, or like beginning language lessons, the songs of Titmice and Carolina Wrens instruct by repeated drill. "The rain in Spain falls mainly in the plain." "Thirty days hath September, / April, June, and November." Or in French *Ou se trouve la gare?* (Where is the train station?). Or Italian, *Quanti anni ha?* (How old is he?).

Repeated rhythm etches in memory: *Thirty days* and *April June* have exactly the same three syllables or amphimacer— long-short-long—as the first three notes of the Wood Thrush's pentatonic song. Start on the minor third, then the tonic, then the diminished seventh. If you play the black keys of a piano, try G-flat down to E-flat, then up to D-flat, which you hold for a couple beats. In the amphimacer rhythm of long-short-long, the first note should be elongated a bit too: *daah-dih-daah*.

Roger Pasquier informs us that among the Bluebirds, Orioles, White-throated Sparrows, and Cardinals, both sexes sing. Among such species as the Mockingbird where the sexes winter and sing in separate areas, once they again pair off on a common area, only the male sings (Pasquier 1977, 113). As for Cardinals, I think I intruded on a domestic conversation in early April, after they had paired. One was perched twenty feet away in a leafless, red-budding maple level with my second-floor bedroom. A male, he looked toward me—and the back of the house where they often nest.

He offered quiet mini-glissandos, which I returned. He uttered some more, which I answered with a Titmouse call. He took off. Possibly he mistook me for his mate out back or stuck inside the house. Adept at dealing with the many mimics among birds, he wasted no time when he discovered the imposture.

Having observed Woodcocks at dusk in early spring in back of my house for twenty years, I don't understand Pasquier's assessment of this species, "where the male has no role in nesting or raising of the young, the song and associated displays are only to attract a female for copulation" (1977, 113). The male certainly chooses an area where a ground nest may have the protection of thorny, creeping vines and other advantages. Perhaps the male Woodcock, with his spectacular display as of a ground army (a guttural squawk) and an air force (with seductive "kissing" sounds in his final "falling leaf" descent) acts only as a real estate agent, recommending a "good" (that is, prickly) neighborhood for the nest.

At any rate, Pasquier notes that, in general, "little work has been done on whether the less frequently heard songs have meanings different from those the bird sings all day" (116). I endeavor to begin filling in this dearth. A corollary of this interest is why certain birds at certain times of year—for example, Chickadees in early March on half-tone different notes—sing more than an hour before sunset. I have observed Robins almost alone in singing in the rain—probably because of the increased availability of worms and the resultant joy. Do Chickadees welcome the oncoming darkness? Or is their competition for *nocturnal* territory—because their two-note song, differentiated by a half tone, sounds very competitive, like Titmices' interminable descending thirds earlier in the day.

After last night's thunderstorms, the late July birds are a bit late waking. I just heard an odd thing, an unusual creaky, croaky sound. Then a few quick, chattery sounds. Then the creaky sound, but more birdlike. Then several truncated whistles punctuated by occasional bursts of two or three notes in succession. It's starting to sound like a Mockingbird or Catbird. I think it's a Mockingbird waking up, warming up its voice. I have heard humans warming their voices on a college visit to Oberlin, one summer at Cornell, and, most lusciously, at the Naples National Library right next to Teatro San Carlo. Through the large, open, second-story window-doors and from across the small park, you could hear the professionals rehearsing trumpet or voice for the evening's performance in another wing of the huge Palazzo Reale.

Back to tonight, coastal New England, late July. Last night's thunderstorms and today's pelting rains have filled the mill-pond where last evening only one rock remained surrounded by water. Along the coastal road, huge puddles. Red-winged Blackbirds mount the utility wires and stare either toward or away from the spectacular sunset. We name the sun's descent with one word, suggesting unity of object and action. Tonight there are vast marches of differing cloud types, some of them hardly more than upward-trailing vapor, some cloudbanks that may run a thousand miles. The Red-wings seem not to know which way to look. They're right: the sunset is a ball of orange fur. It's impossible to imitate Red-wings, although sucking through your teeth makes part of it. Across the street in a bush sings a Song Sparrow, with its three-note initial call varied to two plus two sixteenth notes one half step higher. Following this are incomprehensible, quick triplets, and so on. These sparrows have maculate, streaked chests—their central brown spot worn like a heraldic crest or Olympic medal. Another Song Sparrow across the field uses the conventional same three notes, followed by five-note bursts, first a half step lower. Every year one sings from a couple of trees bordering two fields near my home. Two months ago the song of this homonymous sparrow had the rhythm of a Ping-Pong ball left to bounce on its own, roughly: *duh . . . duh-dih . . . dih-di-di-di . . . didididididi*. It's the rhythm of a mathematical *e*-curve, an exponentially damped oscillation. The Song Sparrow. A Ping-Pong ball, but bright notes sometimes descending, perhaps by quarter tones.

Mall Birds

Late in the day, well after 7 p.m., I drive to a local mall for a printer cartridge. As I open my car door, I hear a glissando call note sounding high. It's also emanating from a height, a bird in the top of some dead branches of an otherwise fine tree. Outlined black against the pre-sunset, the bird switches its tail nervously. Not an Oriole. The tone seems much higher than an Oriole's, though later when I check it on my piano at home, it's perhaps only one tone higher, around a D above middle C. It's more piercing than an Oriole's. With my binoculars, I can see it's a Red-winged Blackbird. And from the bushes I hear the unmistakable meow of a Catbird. Here I am in a mall parking lot, and the familiar birds have nested nearby and found some food. What's that swamp-dwelling Red-wing doing here? Maybe there's a swamp nearby, or maybe this mall used to be a swamp. More likely. Generations of this bird have returned here, even after most of the swamp was filled. I know the same species return to the exact same tree or grove or field along my road, year after year.

How long has this continued? The life of a specific tree sets that limit, a couple of hundred years at the very most. Two trees aged two centuries stand along this road, but most others are no older than a half century. One of the ancient trees rises nearly a hundred feet, an oak. I do hear birds as I pass it, but no specific ones. What if the same species has returned to the same area for a millennium? Keats speculates as much, not about location but about his nightingale's song: "The voice I hear this passing night was heard / In ancient days by emperor and clown." Perhaps even, he continues, the biblical bird referred to in the Book of Ruth.

The next morning, after my mall visit, as I walk along my road at around 7:30 (late for birds), I hear a syncopated, offbeat (not downbeat) descending minor third, the second and third note the same, and twice as fast as the first: *uhdee dih-dih* (again, the *uh* is a rest). Repeated. Then a fainter one up a half step. Dueling Chickadees. They continue, the one sticking to his tones, the other echoic, but in a different key. Together it's a canon. Lovely, too.

Having passed the consort of Chickadees, I hear out beyond two stone walls a call I haven't heard this year until now. Like the Chickadee battle song, this one is syncopated, with a down-beat rest followed by an eighth note, down one full tone to a quarter note, both repeated once, ending on the fifth note, the same it began on.

This is a simple series. I didn't know what bird it was until after I mimicked it, and it continued by glissandoing those two notes and following with a Robin's burble.

One perfect day in late July, I hear a racket at the woodsy end of my house. The noise is a constant, fairly high-pitched cheeping by more than one bird. When I reach the trees, I hear this cheeping near the ground, in the bushes, and then suddenly overhead with some accompanying flitting. Looking up, I see a little bird coming closer, descending from higher in the tree over me. It lands on a branch maybe eight feet away and fans open its tail to draw attention. This is clearly an attempt to distract me from nestlings nearby. It's working. But I do see another bird flitting behind it, maybe a yard distant. It's making the same cheeping, but it has streaks all down its breast—the young of the species, as clearly delineated as a prep school uniform. These are little Chipping Sparrows, with bright streaks over their eyes and forked tails, good for distracting . . . me. Here's a two-week-old creature—at most—following its parent who is doing its best to fend off the myriad threats. I am one, so I retreat back to my porch.

In August near the remnants of woods and a shack by a new mall twenty-five miles from New York City, I hear Crows and Goldfinches. I see a Mockingbird. They have adapted—or have they just not left yet? The diner has been open just a month. Today is muggy; thundershowers expected. The abundant thistle may cause celebratory flight songs from the Goldfinch. Surely Robins sing in the rain anticipating and celebrating the abundance of their vermiform repast. I can't remember hearing such a wealth of Goldfinch vocalization. But that is the way it is with birdsong. You are entirely unaware of it until you associate it with its specific singer. Once you do, you notice it all the time (unless it's a stranger just passin' through).

Homo Homini Lupus

The bird world is harsh. If "man is the wolf of man," birds are surely the wolves of birds: *Avis avibus lupus.* I have seen a small Hawk destroy a nesting Robin's day, a bird no larger than the Robin itself. A Sparrow Hawk or Kestrel came in no more than five feet in altitude, directly into a spruce tree Robin's nest and left with one of its young. It happened fast. There was no air-raid warning or ambulance siren. Not even the usual bird warnings, the cheeping and kecking. The Robins were out foraging. Such a small threat, not my image of hostility at all. I'd expect trouble from a large circling Hawk, but a little one? Believe it.

Since that little encounter twenty feet from my back door, I view every bird encounter differently. Now I know why you commonly see Purple Grackles or Starlings chasing Crows. Or Sparrows chasing a Blue Jay. Crows chasing Red-tailed Hawks. Any smaller bird chasing any larger. It's not a game. Professional tracker Jon Young explains most bird language, including flight patterns, in terms of survival. The major exceptions to this warfare of bird eating bird are fish eaters such as Osprey and Bald Eagle. Shakespeare compares his most aggressive (and undemocratic) hero Coriolanus to the Osprey: "I think he'll be to Rome / As is the osprey to the fish, who takes it / By sovereignty of nature" (*Coriolanus,* 4.7.33). (I'm sure the fish would chase them if they could.)

One afternoon in midsummer I hear the high in-sucking-between-teeth sound of an Osprey maybe a hundred yards over me. Straining to look up while avoiding the sun, I see what looks like a smaller bird chasing an Osprey. As I watch, the Osprey seems first to gain on its pursuer; then the pursuer seems to lose altitude fast. Now I realize it's something the Fish Hawk has dropped. Thump, it lands maybe thirty feet away, luckily just before the woods and bushes. Hurrying to the spot, I find the tail half of a fish, maybe six inches long, five wide. The Osprey's talons have ripped it in half. Possibly the fish struggled and the Osprey gripped harder. The upshot is an Osprey-style filet, jagged and raw-fleshed.

Early on the first day of August I hear a descending minor third, *dih-dah-dah*. I have not heard this for a while, yet its timbre is the now familiar Carolina Wren's.

I wonder why it has altered its call. Perhaps because it was cool last night, in the fifties. Perhaps because of the time of year and the sun angle (it's just before dawn). An hour later when I walk down the road, I hear something on the wing over an open field by some camps. I get out my binoculars but can pick out only a grayish blur. When another lands in the top of a small nearby tree, I see to my surprise it's a Goldfinch. Birds on the wing so often look gray and white from below, even Goldfinches. Its inflight call sounds like four or five descending half tones or even quarter tones. The first and second of these seem syncopated, with the second accented. In other words, there seems to be a rest before the first note. Or consider a series of five notes, quintuplets, the first a rest.

Now at the beginning of August I hear the characteristic eponymous song of the Chickadee. Not primarily a spring song, we tend to hear *chick-a-dee-dee* in fall or winter, except for Barry MacKay who hears it year-round (2001, 27). Still, he notes—no pun intended—that the breeding song of the Chickadee is the two long notes so familiar, so often confused with the Phoebe (Mathews 1910, 219). Written an octave lower, they are:

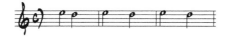

I first recall these from perhaps third grade. Though I may have heard them even earlier in Maine. Schuyler Mathews is eloquent on the beauty of these two notes.

> Poor tuneless Phoebe is intellectually incapable of such a
> perfectly musical bit as this. Mr. Cheney says of these two
> notes, "never were purer tones heard on earth." Indeed,
> few small birds whistle their songs as clearly, and separate
> the tones by such lucid intervals. (219)

He continues with praise of the Chickadee's musicianship, "he knows the value of a well-sustained half-note, another point which should be scored in the little musician's favor," in contrast to the Canary's "hemi-demi-semi-quavers every tone within the compass of an octave" (219). But the human response to this consummate Chickadee beauty varies. MacKay is not convinced, feeling the two notes "a sad, little two-note call that, to my ear, sounds *sooo-sadd*" (2001, 28).

Usually very accurate, Mathews in his *Field Book of Wild Birds and Their Music* makes occasional unwarranted cultural assumptions that interfere with his observations. For example, he states flat-out that "the Cowbird has no song; his nearest approach to music is a sort of guttural murmuring, why should he have either? The bird has no song—no mate to call. He is a polygamist, a bird of no principles, a 'low-down' character" (1910, 53). To the contrary, once in mid-April I heard a lovely call reversing the Goldfinch flight song by rising one tone, *deee-dihdih*. A minute later, having drawn me away from my porch and feeder, the Cowbird appeared at the top of a juniper with his grayer mate, who flew to my feeder, there devouring seeds. They were just back from migration, and they were hungry. Their talk was somewhat like a Starling's.

So tricky is interspecies communication. We project, say, the "mourning" on the perfectly contented call of the Mourning Dove. We bring our associations to the songs. For me, there is nothing more beautiful than the White-throated Sparrow's rising minor third and triplet. They take me back immediately to the rural Maine of my childhood, the piney woods. For others, they may speak of isolation and desolation. Similarly with the Wood Thrush. The first one I recall hearing was in the Berkshires, in Beckett, Massachusetts.

Birds are tied to the solar calendar as well as to climate; for example, the English Robin's so-called spring song begins in late December. This we learn in David Lack's *The Life of the Robin:*

> The spring song of the robin starts near the end of December and continues until about the middle of June. The autumn song, which is thinner and less rich, is first heard in late July from some of the young birds, the adult robins starting about a fortnight later. It is continued throughout the autumn, but is rather feeble in early winter, until the spring song suddenly starts again in late December. The above dates refer to South Devon and are rather different in other parts of England. (1965, 34)

Now on a hazy day in early August I hear several "new" songs or adaptations of earlier ones. I can't say which of these are fall (winter) songs. I do hear a rising variation of the Song Sparrow, whereas in spring the one on my road, for years, has always descended with its increasingly rapid, damped Ping-Pong-ball rhythm after its three piping call notes. Here's his rising variant:

I hear nearby another rising interval, starting on the same note but "thinner." I see a Warbler. Back home, I hear "My Momma done tol' me."

I can tell from the timbre it's a Carolina Wren. When I imitate it, it makes a descending *shree* almost like a Blackbird. This

I cannot imitate. I have never heard such from a Carolina Wren. Could it be, as Mr. Lack says, the young of the species anticipating a fall song, or even identifying its brood?

With all this birdtalk, might I be "the man who thought he was a bird"? This is an ancient medical case about which noted physicians and philosophers propounded cures. In his work *On Chronic Diseases* or *Tardae Passiones,* Caelius Aurelianus lists many types of *mania* or insanity:

> Sic denique furens alius se passerem existimavit, alius gallum gallinaceum, alius fictile, alius laterem, alius deum, alius oratorem, alius tragoedum vel comoedum, alius stipulam ferens mundi se centrum [censum S sceptrum Haller] tenere. (1950, 540)

My translation: So some maniac thinks himself a sparrow, another a rooster in a henhouse, another a pot, another a brick, one a god, another a tragic or comic actor, even a blade of grass bearing the weight of the universe.

No, not a Sparrow. When I fly in my dreams, I am rather cumbersome in flight, more like a Heron or Cormorant with wet wings. Birds and flight appeal to our irrational brethren, as well as our wisest (see "Avitherapy," p. 69). As for a rooster in a henhouse, doesn't every adolescent male feel that?

The first week of school crossing the sunny parking lot, I hear a piped, staccato descending major third. Turning around, I see a Blue Jay at the top of a twenty-five-foot tree. I've heard just the first staccato note for a week or two in early spring, but this seems to be the fall variation. A week later as I leave campus after too many classes, I hear the same call and mention it to a student who is standing nearby, her two children asleep in the car. I imitate the call and remark, "That's a Blue Jay." She's surprised, because it's not their usual jeer. After I echo it three or four times, worried I might wake the kids, the Jay swoops from a grove of scrub oak some fifty yards away. It crosses in front of us some forty feet away and maybe thirty feet up, and then it lands on a tall lamppost. It has checked us out. What did it conclude? I am no threat to its feeding territory? I am not one of its brood? I am not a bona fide Jay? Probably all of the above.

By early October, the remaining birds—Blue Jays, Carolina Wrens, Chickadees, some Catbirds (those that migrate may have genetically inherited a gene for migration!), some Sparrows (the White-throated, but not the Song)—offer their spare morning songs. But after a miserable northeaster, after a night in the forties, the day opens with bright sunlight around seven o'clock, and a bird scolds with its *tchk, tchk*—probably at a neighbor's cat. In the trees to our east that catch the first sunlight, a Carolina Wren offers its triplets. A Jay pipes its resonant, overtone-rich call. They appear to be greeting the sun; they appear to be cheered. Surely we would not presume too much to call this, in fact, *singing*.

A week later, rounding the bend amid yellowing and rusting leaves, a haze of bright sun glows in the lifting fog over the rattling cornstalks. It's about 8 a.m., an hour after sunrise. Birds chatter and whistle in the remaining leaves and among the brittle cornstalks. They seem to be celebrating. If complex human beings feel buoyed by sunlight and even by foggy auras, we should not doubt that the birds in their immediacy and—is it simplicity?—do, too.

Later in the week, now mid-October. Another foggy morning, dew on everything as after rain. The brook, though,

Angeline Brook, is pretty dry. I hear a trickle. Then two distinct, piped notes behind me. A descending minor third, up at the end of my whistling range, from the F above high C (the one above middle C—the first note children learn on the piano). I whistle it back, returning his call. Again he pipes the falling minor third, the same notes as one of the Carolina Wren calls described earlier (see "The Year of the Oriole"). But this doesn't sound like that hearty wren. I hear another similar call a couple of hundred yards away. That's his correspondent. I am merely an interloper. I speed up my walk back to where the minor third first issued. A Catbird flits across the road. Could it be? After a couple of warning sounds so prevalent in fall, there's the piped descending call. So the Catbird makes one, too, in its own timbre. Like the Titmouse and the Robin and the Carolina Wren and a legion of other birds, the Catbird sings minor thirds—but more often in fall than in spring.

As the leaves in the bright October light yellow and fall, I suspect one of the prevalent uses of birdsong diminishes. That use is deception or wily invitation to pursuit. Reflecting on the Catbird's minor third, so like several other species' calls, I wonder if this is mimicry or serendipitous duplication. What would be the advantage of mimicry or imitation? This question must occur to an apelike creature, particularly one writing a book on mimicking birdtalk.

Clearly, birdcalls have a range of survival values. Avians, who stayed in and above the trees, must have an entirely different sense of space from us creatures who descended. Most avians' sense of flight distance, the distance at which they decide to flee or remain, is much greater than ours. Most any bird can, within five seconds of taking off, be far enough away—a few hundred yards—that we sapiens would consider it outside our lot or property or space. Any novice birder has had the experience of walking too close to, say, an American Heron—perhaps seventy yards. We were in his space before he was in ours.

Birdcalls bridge such distances for all species necessities, such as feeding, mating, and defense against predation. Birdsong, especially deceptive sound, helps birds to "hide" in broad daylight. Even nondeceptive song must share, like plumage, the purpose of survival. When a male Cardinal alights in a tree and boasts—or "yells" or "sings"—his series of descending glissandos, he may well be maneuvering to deceive predators and egg-eaters such as Sparrow Hawks and Blue Jays. His song works in tandem with his plumage to attract attention to himself. He is providing potential predators a locus of attention other than his nearby nest. So even his most earnest, boastful, upfront, "sincere" song has its deceptive intent. The implicit vanity of drawing attention to oneself seems to dominate both avians and simians. Our two species share this. A Cockatoo I know will not screech obstreperously as long as he's the center of attention. We sapiens may also draw attention, or we may drive fancy-plumaged cars or wear screechy, "loud" clothes, or we may even write.

Now, in late October, I hear among yellow leaves a high slurred note descending perhaps a half tone, perhaps a whole. Lots of species have a similar call. Orioles, for instance, have a flight song just like it but an octave lower. Goldfinches make a similar call, though their timbre is thinner. Catbirds make one just like it. Searching out the source of the slurred notes, I look through binoculars across a hundred-yard field: A medium-sized hawk rises from the trees beyond. I don't hear the slurred notes again. Could it have been the hawk? Probably. The Osprey, for instance, makes almost exactly this call, but a bit higher and more piercing. Earlier I had passed within a rod of a couple of dozen Herring Gulls who ignored me in the twenty-knot breeze until I answered their high *scree.* I sucked in between my teeth, and they leapt into the air. I had ineptly called like a Hawk, probably. Sounds we are deaf to, and cries on the upper border of the human ear, commonly punctuate birdspeak. It is pretty hard to get it right, but if you don't, they flee. Birds are sensitive to mispronunciation, even more sensitive than the French. Birds have reason to be, because "misspeak" is often their only warning of a predator, rather like passwords in wartime.

Blue Jay

Song, Assault, or Mockery?

*W*hat do we hear when we listen to birds? What for us may seem a melodic interval may be a desperate and distressed plea. What passes as hostile assault between two Titmice (even one precipitating avian "battery" as well— pecking tailfeathers and so on) may sound like a musical fugue. Even to call bird sounds "song" is to project a certain relatively narrow range of human activity. For birds, what we call their "songs" are more often curses and threats, defiance or concession, arrests and apprehensions, or even weather bulletins and realty leases. We simians do not tend to "sing" upon being stopped for speeding; a bird would. Nor do we "sing" when a deadly enemy approaches our home to steal and possibly eat our offspring. Or if we do sing on such occasions, we are doubtless onstage in a Greek tragedy or grand opera.

We should think a bit before we apply human behavior to birds. One way is to reverse the bird studies. Would such assessments of birds make sense if they were about us? For instance, ornithologists classify most avian behavior under territory, mating, and feeding. The brave scientists add migration, although because it is not well understood—often migratory and non- migratory branches of the same species share territory—very few studies link migration and "song." Now, if we reverse these studies, we say that most human behavior falls under social and sexual life, finding a place to live, and food. The brave sociologist or psychologist adds travel or *migration*—a fairly accurate term for American retirees flocking to Florida or Arizona.

Almost all human behavior can be reduced to such elemental necessities. Clothes and jewelry? Mating behavior. Cars, roads, and fences? Territorial determination. Agriculture, restaurants, and TV chefs? Feeding. I suppose one could subdivide the whole world economy and every separate culture into such basic categories. I feel, as a human, that such reductive classification, commonly practiced in field studies of birdsong—and arguably practiced in certain branches of anthropology and especially psychology (Maslow's "hierarchy of needs"?)—such classifying does not adequately assess *my* communications and expressions.

Mysteries abound in the songs, or the curses and threats, of birds. In his chapter "Emotions" in *The Minds of Birds,* Alexander Skutch attributes human emotions to bird vocalizations. His approach inherently questions the usual scientific assessment of bird-speak for its function or use—either to establish a mating or feeding territory, to attract (or, logically, to repel) a mate, to signal the male's readiness to take over egg incubation, and so on. Skutch notes that we human observers find or project our own emotions:

> It is remarkable how often the sounds that birds make suggest the emotions that we might feel in similar circumstances: soft notes like lullabies while calmly warming their eggs or nestlings; mournful cries while helplessly watching an intruder at their nests; harsh or

grating sounds while threatening or attacking an enemy; sharp, castanet-like clacking of the bill while trying to intimidate a rival or an interloper. Birds so frequently respond to events in tones such as we might use that we suspect their emotions are similar to our own. Some birds, however, appear to lack notes appropriate for harassing occasions. (1996, 42)

That is, they have only beautiful songs! Ah, to be hemmed in and confined to beauty—to be relegated, like Midas to gold, to speak only poetry.

Skutch notes a function of birdsong identified by only a few other ornithologists, the propensity of a nesting bird such as a Flycatcher to "voice a little nest song as she passed through the doorway in the side of her bulky roofed nest and snuggled down on her two speckled eggs. Sometimes she twittered softly in the midst of a session of incubation and at intervals she called more loudly and was answered by her mate" (Skutch 1996, 37). One easy way to assess birdspeak is to differentiate the bird's tone from what human speech would be in an analogous situation. For territorial calls, we may expect the human male to beat his breast, apelike, or to sing and play a rock song on amplified guitar. Or possibly the territorial call is closer to a real estate closing, with two sets of lawyers, bank reps, new owner, and so on. Somehow, the Wood Thrush and the Catbird win this comparison either way.

Reversing the analogy, what if humans could only sing beautifully, no matter the occasion? We would be living in Verdi or Donizetti or Gershwin. And who can doubt that after their silence of a cold autumn evening, birds chatter more at the advent of sunlight? No matter their various songs; they all may be singing, really expressing overflowing emotion, as if they knew "*O sole mio*":

> *Che bella cosa 'e ne jurnata 'e sole,*
> *'N'aria serena dopo 'na tempesta!*
> *Per l'aria fresca pare gia 'na festa . . .*
> *Che bella cosa 'e ne jurnata 'e sole!*
>
> (Paliotti 1992, 126)

Just over a century ago in Naples, Giovanni Capurro wrote, "What a beautiful thing it is, the daylight, the sun / In the peaceful air after a storm, / The air so clear it's like a party already, / What a beautiful thing, the sunshine" (my translation). Once Di Capua set this to music, the birds I heard this October morning could well agree. I suspect that much bird-speak is also weather-related, often seasonal but also daily or circadian.

By the first week of November, the birds are flocking, gathering among the leafless branches "where yellow leaves or none or few do hang," as Shakespeare put it. There are Jays in abundance, a dozen here in the scrub oaks, and Robins as well, several dozen. I try out the piped descending minor third that worked a couple of weeks ago; it called a lone Jay across the parking lot. But this morning none responds. They hang together hopping among the bare branches. They do seem agitated; maybe it's their version of Zugunruhe, premigration restlessness. They vocalize in a kind of *chee,* also imitable by in sucking, but lower than the Osprey. Today Robins, like the Jays, ignore my usual, burbling Robin salute. Nor do they hail each other in their characteristic calls.

I conclude that flocking and migratory birds have no use

for their characteristic songs. When their behavior changes, their language changes. Their usual calls imply certain significant spaces between them; such songs might be compared to operatic passages, inappropriate in a crowded subway or taxi. Their usual songs also suggest greater individuation than during migration. Remarkably, they no longer regard calls that were answered just two weeks earlier. A comparison in human behavior might be joining the army and its attendant haircut and uniform. Birdsong functions as do civilian hair and attire, as well as property deeds.

Many birds also mimic. At least, we hear their birdtalk as imitation. Mockingbirds are named for this ability, but they are not alone in this. Blue Jays can imitate almost as well, although they don't do it from the tops of trees in rapid succession. While Mockers appear to perform gratis—thespians incarnate (or pennate)—Jays use mimicry to lurk and then to surprise their prey. Who is the better mimic, the one everyone knows is imitating, or the one we don't notice? If the Mockingbird delights like an actor or even a parodist, the Blue Jay, like a double agent, prefers the shadows. He talks clandestinely in several bird dialects.

One mid-February morning in a dusting of snow, I heard an apparent Carolina Wren repeat a grace note and descending minor third. Looking for the bird typically halfway up some leafless bush, I could not find it. After several exchanges, I finally saw a Blue Jay low to the ground. Once I saw it, it made a guttural *caw* like a small Crow. In *Bird Sounds* Barry MacKay comments on such Jay ability, and on the nature of mimicry:

> Blue Jays have a call that is similar to that of Red-Tailed
> Hawks. Whether or not this derives from true mimicry,

the cry so closely resembles that of the raptor that some birders do not assume that a hawk's call actually comes from a hawk until they see the hawk make the call.

Blue Jays, then, make a wide variety of non-Jay calls. Many of these apparently deceive, like duck hunters *quack*ing from a blind (2001, 46).

When are similar songs true mimicry? The story of bird mimicry does not begin and end with deception and parody. This book started with a wide variety of birds making similar sounds and intervals—descending minor thirds, ascending fourths, glissandos. Would it be too much to call this the lingua franca of the species? It may well turn out that certain sounds are universally recognized among birds. Surely their warnings, the fricative *chuck-chucks* and in-sucking lisps, are clear even to simians. What then can they make of each others' descending minor thirds? Do they function like stone walls in New England? Or like answering the phone? They do seem to hold a function quite beyond any imitation or mockery.

Magpie

Auspices

Roman and Native American

We humans have always turned to birds, to their flight patterns and their songs, for signs and portents. Ancient and prehistoric people observed birds to prepare for the future, as much as seeking among the counseling stars and voices in the wind. Their "future" could have been the next hour, the day ahead, or a lifetime. By Roman times, just at the start of the Empire, one member of the college of augurs, Marcus Tullius Cicero, lamented how the last ten years had seen the decline of consultation of bird signs, the auspices. In *De Divinatione,* Cicero's brother Quintus contrasts Roman neglect with the greater piety in the outlying provinces of Pamphylia, Pisidia, Lycia, and especially Cilicia, where the Cicero brothers were Roman governors.

> *Quae quidem nunc a Romanis auguribus ignorantur . . . a Cilicibus, Pamphyliis, Pisidis, Lyciis tenentur.*
>
> (CICERO 1954, 252)

While Cicero governed Cilicia with his brother's assistance, they met and befriended King Deiotarus of Lesser Armenia and Gallograecia. Making the argument for auspices, Cicero's brother cites this ruler from the mountain regions. King Deiotarus always consulted the auspices—the avian augurs—before and even during a trip. Several times he stopped his travels because of the auspices. Once he returned home because of the flight of an eagle. The next day, the room that he was scheduled to stay in collapsed.

Both birdsong and bird flight have been considered premonitory. Perhaps Shakespeare depends upon such ideas when, for instance, Juliet and Romeo debate whether it is the Nightingale or the Lark who sings and presages Romeo's arrest (*Romeo and Juliet,* 3.5).

Juliet: It was the nightingale, and not the lark,
        That pierc'd the fearful hollow of thine ear;
        Nightly she sings on yond pomegranate-tree:
        Believe me, love, it was the nightingale.

Romeo: It was the lark, the herald of the morn,
        No nightingale: look, love, what envious streaks
        Do lace the severing clouds in yonder east:

Then Romeo accedes to Juliet's desire that he stay. With bravado, he says he is content to be captured, even to die, if that is her will. Like Petruchio, who misnames the irksome Kate a fair and modest maid, Romeo misnames the bird, "Nor that is not the lark, whose notes do beat / The vaulty heaven so high above our heads." But such misidentification could be fatal for Romeo, two lines later, "Come, death, and welcome!" Then Juliet awakens to their danger, foreboded by the correct song identification.

> It is the lark that sings so out of tune,
> Straining harsh discords and unpleasing sharps.
> Some say the lark makes sweet division;
> This doth not so, for she divideth us.
>
> (3.5.26FF)

Juliet's "field I.D." of the lark is well expressed, though she might say the same for Messiaen. The beautiful song of the lark here warns the lovers, as elsewhere in the canon, in *MacBeth,* the freedom-loving Purple Martin ("martlet") greets the king entering the castle where he will die.

> This guest of summer,
> The temple-haunting martlet, does approve
> By his loved mansionry that the heavens' breath
> Smells wooingly here.
>
> (1.6.3–6)

To the naive observer of feather headdresses, dances, and birdcalls, aboriginal cultures are most engaged in the cross-species encounter I discuss. Birds are entwined in so much tribal lore. Although Native American encyclopedias often do not include an entry for "bird," when they do, it is often a translation of a family name. When I look up bird species, such as Crow, Eagle, and Owl in *The Encyclopedia of Native American Religions,* I find more proper names—David W. Owl, a Cherokee Baptist minister whose parish was among the Seneca on the Cattaraugus Reservation; Adam Fortunate Eagle, one of the activist occupiers of Alcatraz Island for nineteen months beginning in 1969 (Hirschfelder 1992, 205; Hoxie 1996, 13). Clearly, birds are so central to native peoples that tribal subdivisions, moieties, or "houses"—and even families—take on bird names. A recent biography follows Zintkala Nuni, also known as *The Lost Bird of Wounded Knee* (Flood 1995).

In William Fenton's article "North Iroquoian Culture Patterns," a Seventeenth-Century French document records copies of native—probably Seneca—pictographs of family totems of those in a war party. Each beast holds a war club, ax, or knife. After the Turtle, Wolf, Bear, Beaver, Deer, and Wild Potato (!) come the Large Snipe, Small Snipe, and Hawk (with club, knife, and ax, respectively) (Fenton 1978, 299). Councils are represented by totem animals with mouths open in apparent conversation—a multiplication of our theme of cross-species communiqués. Some family names reflect such species mixes—for example, Henry and Mary Crow Dog (Sturtevant 1994, 990–94).

But ethnomusicology comes to our rescue. William Powers's study of Lakota sacred language establishes the primacy of Native Americans in cross-species communication. One of the sacred Oglala languages can be used with nonhumans.

> It is believed that a number of animals and birds can communicate with common people and medicine men and that frequently these nonhuman species speak in Lakota. The communication between humans and nonhumans occurs most frequently during the vision Quest. Under ritual conditions, virtually every species is capable of speaking to the medicine man in Lakota—buffalo, deer, elk, coyotes, wolves, snakes, and so forth.
>
> Sometimes the animal or bird is capable of signaling the medicine men in the species's natural utterances, for

example, the howl of a wolf, the bark of a coyote, the hoot of an owl, and the crowing of a cock, all of which are interpreted as omens of death particularly if the sounds are heard during an inappropriate time of day or night for that animal or bird to be active. Most often, communication between people and animals occurs in the vision Quest, Sweat Lodge, or in the Yuwipi. (1986, 26)

Interesting how time of day, chronobiology, factors into the meaning of animal sounds. Messiaen would heartily agree, as we see later (see "North Atlantic Birds").

I would argue that many of the difficulties early musicologists found in notating Native American songs crop up in birdsong as well, from quarter tones and other surprising intervals to glissandos and slurred effects. In one early, elsewhere wrongheaded attempt, Frederick Burton makes at least one point of agreement:

There is much vagueness in the Indian's frequent slurring from one note to another, the intervals are often, to say the least, unexpected, their scale relationship hard to determine until long familiarity with both style of songs and performance enables one to listen in much the same spirit with which the song is sung. (1969, 22)

Birdtalkers must concentrate and adapt their ears much as ethnomusicologists learned to in their field studies. I have only started to hear quarter tones after years of trying to notate avian intervals, years of finding my transcriptions inadequate.

In William Powers's chapter "Naming the Sacred," we

find that the Lakota make at least four distinctions to our
two regarding Golden Eagles and Bald Eagles, adding specific names for immature Bald and immature Golden (1986,
148–52). The Lakota also associate certain birds with certain
hunted animals, for example, the Swallow and the black-tailed
deer, the Crow and the white-tailed deer. At the same time,
certain birds are harbingers of specific seasons—the Crow, of
spring; the Swallow, of fall. The Magpie announces winter, the
Meadowlark, summer.

Rhode Island's founder, Roger Williams, in his *Key into the
Language of America,* records some of the Narragansett language for birds in 1643. In his chapter "Of Fowle" after the
word for Eagle (*wompissacuk*), he records those for Turkeys
(*neyhom*), Partridges, Heathcocks, and especially Blackbirds
(*chogan, euck*):

> Of this sort there be millions, which are great devour-
> ers of the *Indian* corne as soon as it appears out of the
> ground; Unto this sort of Birds, especially may the mysti-
> call Fowles, the Divells be well resembled (and so it pleas-
> eth the Lord Jesus himselfe to observe, *Matth.13.* which
> mystical Fowle follow the sowing of the Word, and pick it
> up from loose and carelesse heareres, as these Black-birds
> follow the material feed. (1971, 89)

Williams records that the Narragansetts would post their
biggest children in "little watch-houses in the middle of
their fields . . . early in the morning [to] prevent the Birds."
But they would not harm Crows (*kaukont tuock*) because
of lore:

Scarce one *Native* amongst an hundred wil kil them
because they have a tradition, that the Crow brought
them at first *Indian* Graine of Corne in one Eare, and an
Indian or *French* Beane in another, from the Great God
Kautantouwits field in the Southwest from whence they
hold came all their Corne and Beanes. (90)

The Narragansett word for Goose is onomatopoetic, *honck* or,
plural, *honckock*. They loved hunting ducks and looked forward
to using European guns for that purpose.

On other birds besides Crows, similar unwritten stories
enter into cross-cultural—or in our case, cross-species—
encounters. In his book *The California Indians,* Heizer tells
of ceremonial sacrifice of Condors and Eagles, and the use
of bird parts in many ceremonies (1971, 124). Note that the
most feared bird of all, among several California tribes, is the
Western Meadowlark. I would hazard this fear derives from
its voice and song—beautiful, varied, and penetrating. Even if
one does not believe in spirits, the Meadowlark seems to bear
a message from another world. For us, that world is Birdworld.

Like good Nineteenth-Century naturalists, the
Narragansetts hunted and ate most all fowl, even our mis-
named, onomastic Buzzards Bay Cormorants (*kitsuog*). Roger
Williams tells us their method: "These they take in the night
time, where they are asleepe on rocks, off at Sea, and bring in
at break of day great store of them" (1971, 91). Among other
bird lore, Williams reports the Narragansett bird name *sachem:*

A little Bird about the bignesse of a swallow, or lesse, to
which the *Indians* give that name, because of its *Sachim*
or Princelike courage and Command over greater Birds,

that a man shall often see this small Bird pursue and put
to flight the Crow, and other birds farre bigger then it
selfe. (92)

In our time the birds most likely to be chasing Crows are
Purple Martins and others in the Blackbird family. I have seen
them also chase Red-tailed Hawks. Williams quotes the word-
phrase *sowwanakitauwaw* for "they are going southward," when
the "Geese and other Fowle at the approach of winter betake
themselves in admirable Order and discerning their Course
even all the night long" (92).

As for songbirds, they are largely beneath Williams's sur-
vivalist notice: "There are abundance of singing Birds whose
names I have little as yet inquired after" (92). However, he does
conclude, as I have *pace* St. Francis, that birds teach *us* a kind of
gratitude or "natural piety":

> How sweetly doe all the severall sorts of Heavenly Birds, in
> all Coasts of the World, preach unto Men the prayse of their
> Makers Wisdome, Power, and Goodnesse, who feedes them
> and their young ones Summer and Winter . . . although
> they neither sow nor reape, nor gather into Barnes. (93)

Williams concludes his chapter on birds as he does his others,
with a hymn, this one in part:

> *If man provide eke for his Birds,*
> *In Yard, in Coops, in Cage.*
> *And each Bird spends in songs and tunes,*
> *His little time and Age!*

Mockingbird

The Supreme Mocker

*E*agles, wings sculpted open, perch atop the two flagpoles in front of the U.S. Supreme Court. Ben Franklin would have preferred a departure from that avian symbol of empire; he nominated the Turkey for national bird. I had never seen a Wild Turkey in flight until this year—two rising from a rural road, a magnificent sight, rather like the takeoff of a C-130.

But back to D.C. On summer days those Supreme Court flagpoles host another avian on top of one eagle: here sits the court jester, a Mockingbird. I whistle back the intervals and appogiaturas, the descending minor thirds, the demiquavers of his grab bag of melody: the Cardinal's descending sixth glissando, the Carolina Wren's "Mexican Hat Dance" theme. No Thrush song this morning, so perhaps he observes some diurnal decorum. Wood Thrushes usually complete their full song in the evening. Pentatonic, it is the same scale as the blues. Many bird notes, like blue notes, fall between two diatonic notes. When Thrushes sing the blues, they sound like Stan Getz were he to play the flute. The notes float and sail.

My East Capitol Street Mocker leaps vertically in the air above his eagle perch, helicopters for a couple of seconds and displays his white wing stripes, then settles back on the finial eagle's wing. Though feathered of metal, the eagle and he are about the same size. The Mocker's look-ma-no-hands wing display completes our introduction: He has warned me. I'm in his space. Has he noticed my broad-brimmed hat, and beneath it a giant nestling, no master of flight? His "wordhoard," as the Anglo-Saxons called it, comprises a vast anthology of other birds' calls, done in his own timbre, sometimes with his own added microtones. It is relatively easy to whistle back to Mockingbirds; if you can't repeat one of their stolen songs, try another. The one I have called the Cardinal's descending sixth usually has a blue note or microtone grace note so it sounds like a triplet.

Most Mockingbirds have a couple of dozen songs within the range of human—read *my*—whistling. (Contrast Red-winged Blackbirds, with their lovely filigree of ultrasonic squeaks.)

Staying out of that Mockingbird's territory is not a problem since I am wingless and moreover settled in the air-conditioned Folger Shakespeare Library. My whole training as an academic has been territorial, to keep off others' academic turf. So I know the message, although I have trespassed often—into journalism for the *New York Times Magazine,* an Oscar-nominated film, folklore, Italian Risorgimento history, even the history of science and medicine, and of course American literature, though my training is in Renaissance English. Journals and conference directors have often greeted me with the academic equivalent of "Keep out of my tree!" Only my talented, multivocal, polysemous academic mentors manage to sneak across disciplines with ease.

On a visit to Monticello, as I leave the confines of the all-weather passage and the south pavilion, I hear another Mocker. Less multivocal than the Supreme Mocker, this one has the whole giant garden and the orchard beyond to himself. The heat of the day, high nineties, has peaked and passed. This Mockingbird does not mimic snippets of Thrush. Mostly he pipes a minor third, two notes in rapid succession, like a Titmouse only more nervous.

I echo back my version of his song. He adds another three or four riffs. I add my version of them, too. Where is he? Something shoots out of a fruit tree, and as it swerves I see the mocking stripes. He measures out his area, or he leaves it to me. Have I won the field? Hard to say since it is no longer mating season.

My stay at the Folger over, I pause at a rest stop on the New Jersey Pike. Cars unload their passengers, swimming in August heat. I take the plunge. The desolate pavement exudes waves of hot air, the trash bins fill with cups from soft drinks that have sustained life between Hightstown and New Brunswick. Two youths sleep in the backseat of the maroon car parked, windows open, next to me. My dulled mind lifts, surprised by a wordy Mocker-locutor who hails from a maple tree bordering a field that abuts the gas plaza. I try to whistle back, but my lips are parched. I sound like a child learning to whistle. A passing couple look at me. Should I explain? "Oh, I really *can* whistle, I can too. I do it all the time, mainly to Robins, Titmice, Cardinals, Orioles, and Mockingbirds. There's one over in that grove. Hear that? No, there really *is* one. I'm not bananas." I let them pass by unenlightened—and myself unexposed. I manage a couple of phrases in Mockerese—or mockeries. When I return from Roy Rogers with my coffee, the bird has flown.

Even here amid the asphalt smell and the gas pumps, the

paper cups and brimming bins, nature takes its late-Twentieth-Century stand. The birdsong, not to mention the "birdalog," transforms the dire necessities—gas, oil, and water for the car, other liquids and facilities for the body. If I were to lead a movement—D.C. makes one think politically—it would be for talking to birds in urban and highway environs which we have created and they do not entirely disdain. Most of what we say to each other in species encounters may be on the order of fashion statements: "Look at me." "No, *me*." But some of it is on the order of much human conversation: "Nice day, 'nt it?" Or George Harrison's "Here Comes the Sun." And "*O sole mio*."

As I get in my car, I notice the fellows are no longer sleeping in the next. They have flown, too. I stand cheered, rested, by this rest stop, more by the bird than all the rest—pun intended. Shakespeare calls the Lark's morningsong "hymns at heaven's gate" (Sonnet 29). I recall also that wonderful passage of throwaway verse late in the *Merchant of Venice* when, awaiting Portia's return, Jessica complains, "I am never merry when I hear sweet music." Lorenzo psychologizes, analyzes.

> The reason is, your spirits are attentive.
> For do but note a wild and wanton herd
> Or race of youthful and unhandled colts . . .
> If they but hear perchance a trumpet sound,
> Or any air of music touch their ears,
> You shall perceive them make a mutual stand,
> Their savage eyes turned to a modest gaze
> By the sweet power of music:
>
> (5.1.70ff.)

The passage goes on to become famous, with its more invidious lines against those who do not appreciate music: "The man that hath no music in himself, / Nor is not moved by concord of sweet sounds, / Is fit for treason, stratagems, and spoils . . . / Let no such man be trusted." Following Lorenzo (or Shakespeare), I urge that those who fail to whistle to birds not be allowed to vote, to drive, or to bear arms. Take your pick. Barring such stipulations, I shall continue to organize the nondenominational Society of St. Francis of the Gas Plazas.

Goldfinch

From Guns to Birdfeeders

The Twentieth Century saw an enormous change in attitudes toward birdlife. The first decades witnessed a burgeoning of ornithological studies but often as an extension of hunting and capture. A reader of the *Auk*'s predecessor, the *Bulletin of the Nuttall Ornithological Club,* for 1882 is greeted with one Dr. Merriam's cheering account of a social occasion in the service of ornithology.

> One rainy afternoon about the middle of July, while the Judge was catching salmon at the famous "Upper Pool" on the Godbout, Mr. Nap. A. Comeau and I climbed a high and densely wooded hill that rises from the western border of the pool, and when near the summit saw a Pine Grosbeak, in the slate and golden plumage, hopping about among the branches of a large Balsam (*Abies balsamea*). I was within twenty feet from the bird, but having only a rifle was unable to secure it. (Merriam 1882, 121)

To "secure" has an ominous tone for the modern reader, for, like his predecessor John James Audubon, Dr. Merriam meant to collect it as a specimen by shooting it. In this case, he missed.

His companion did not, for, following the physician's example, "he has since written me that he has shot several after I left." This serves me as an admonition. Writers on birds inevitably desire to share their often solitary, inevitably hermit-like pursuit. Their readers may not share their proclivities—for instance, killing it to save it.

My own interest, too, can be seen as an intrusion, a mucking up of natural order. When at my best as an avian mimic, I may be like a decoy Puffin placed on an island to attract released nestlings back to a former site. My "answering calls" may confuse the territorial males and the females as well. What if I attracted more females than the real birds? That would be a reproductive and genetic disaster. Fortunately, that has not happened. I seem to attract only males and only at their most territorial.

Why do this? Perhaps it is the monkey or the mocker in me. Perhaps it is the idealist, perhaps the nostalgic. The idealist wants to break barriers between species, the nostalgic wants to connect with nature as in childhood when, according to Wordsworth, we were closer to nature and to God. But most likely I echo birds as a linguist and musician. I have studied languages, European Romance and Slavic, much of my life, although my training is literary. I have spent hours, whole days, some years slogging through various texts that require dictionaries. I enjoy learning what I do not know, especially when veiled in a language I can but partially construe. For many years I have been an amateur jazz musician, even the composer of a few jazz tunes, some based on Wood Thrush and Oriole songs. So this book is a prelexical approach to a dictionary of bird language—confining ourselves to the minority of sounds imitable by us—or really, by me.

Essentially, imitating birdcall notes is like imitating a telephone dial tone. A connection has been made. That is it. In his book on English Robins, David Lack makes this point.

> Analysis of the sound-tracks of bird song has shown that most small birds include in their songs notes of a frequency well above the limits to which the human ear is sensitive, so that a bird's song probably does not sound the same to another bird as it does to us. (1965, 34)

Lack seems to forget this wider range of avian perception when he notes the prevalence of both female and male English Robin singing in the autumn:

> In this season not only the cocks but also about half of the hen robins sing, their song being indistinguishable from that of the cocks. (34)

I am sure that the birds can tell the difference easily. The future of the species depends on it.

It may also depend on not landing in Maurice's backyard. Enthusiastically recounting the great White Goose Festival along the St. Lawrence River, our host Maurice tells of the sky white with Geese and getting his shotgun to provide supper. I always recommend Konrad Lorenz to Goose hunters; Lorenz's *King Solomon's Ring* recounts his experiences with Geese who mate for life, their young bonding with humans should they lose a parent.

Even if you have no interest in extending your parenting role into the avian sphere, more good reasons exist to call birds—or, rather, to return birds' calls. Birds are tasteful conversationalists.

They do not talk about digestion in restaurants, nor about flatulence, though all those seeds and berries must do something. They do not converse about their health problems, about Dr. Bierstein's latest discovery from the blood test and his consequent recommendation of which trendy drug you've seen advertised on TV. Nor do birds talk at great length about sports teams in their sumptuous new stadiums, about the latest contract negotiated by this knuckleballer or that quarterback. The stadiums they evaluate only as nesting sites. The newer ones are often worse, hostile to birds and far from wild fruiting bushes and other feed—except mosquitoes and bugs. Plenty of those, though the electric zappers are eliminating food supply there. Birds' stock portfolios are remarkably thin, although some few species do hoard, and their discussion of investments can be summed up in the seasonal ebb and flow of sunlight. They do chatter a great deal about weather, and rain causes many to sink into silence. Not the Robins, though. Even among humans, the avian ability to discuss weather instead of serious subjects has always been a sign of good breeding. And birds are nothing if not good breeders. The migratory species would probably watch the weather channel before migration if they knew human language. Certainly butterflies should, too. When Hurricane Bob tore through Rhode Island at 110 mph, I saw dozens of Monarchs caught in their migration swept backward at very high speeds.

In fact, I suspect birds have their own weather channel, or at the least, weather reports. Late October I heard around sunrise very rapid sixteenth-note triplets, an ascending minor third: C sharp, D, E: CDECDE CDECDE. Two beats a second, twelve notes a second. This was a familiar Carolina Wren call, but faster than I have heard it. Perhaps the speed reflected the temperature. The astronomer Harlow Shapley once created an accurate thermometer from ants because their speed varied as did the temperature. In the case of ants, they were the opposite of humans, walking faster in the heat, slower in cold. I think birdsong contrarily speeds up in cold weather. The Carolina Wren's late October morning song is essentially a weather report. But below a certain temperature, there are few such reports. Or any other news from birds.

Finally, birds do not bemoan or complain about their work, although they certainly could. They search all day, every day to find food and to avoid predators. They could heap a torrent of abuse on their interlocutors. Swallows surely demonstrate great energy at dusk sweeping across fields for flies and gnats to swallow—pun intended. Robins watch eagerly with their characteristic head tilt, hour upon hour, for those worms. Chickadees gather seeds off trees, Goldfinch thistle seed, off August driveways if necessary. Essentially trapped in the hunter-gatherer stage of cultural evolution, birds spend the long daylight hours, every waking moment, looking for food. At least, the oscine passerines do. Their search for seeds and fruits and bugs must work out about even in calorie intake versus expenditure. And when they're feeding young, they fall behind. Any mother Robin or Cardinal looks frantic and thin about ten days into her twelve-to-fourteen-day nesting cycle. The males look frantic, too, among species where both feed the young. Fish eaters, such as Osprey, Kingfisher, Heron, and Eagle, seem to have slightly more "free time" just to soar. Or maybe that is just my human perception based on our species' prejudice in favor of a protein-rich diet.

Such observations and appreciations grew out of the Twentieth Century, which brought us from Audubon's shoot-and-sketch approach to birdlife. Perhaps the Twenty-First

Century will bring us into a conversational acquaintance with our airborne rivals.

As surely as the Nineteenth Century saw the growth of the capture model of bird study, it also fostered some wonderful "natural piety," to borrow Wordsworth's phrase. For instance, in France, historian and journalist Jules Michelet wrote a lovely book toward the end of his life, a retreat from the *sujet humaine*. He contrasts his book *L'Oiseau* to his predecessor Tousenel's, which is pervaded with the "*vocation militaire du Lorraine*" (military vocation of Lorraine). His own is a book of peace, against hunting, "*en haine de la chasse*" (Michelet 1876, 9). He promises to avoid his mentor's human analogies, his continual personifications, a device common among writers on birds—a fate not entirely avoided by myself.

In rejecting the usual Nineteenth-Century hunt-capture-and-classify approach, Michelet accepts the gender taunt that his approach is dreamy and feminine. That the heart of a woman is mixed in his book, Michelet says, "*Nous acceptons comme un' éloge*" (We accept as praise) (10). In his chapter on song, Michelet writes a vivid hymn to the birds:

> Winged voices, voices of fire, voices of angels, emanations of an intense life superior to ours, of a life voyaging and mobile, which bequeaths to the laborer tied to his furrow thoughts more serene and a dream of liberty. (256)

He continues in this elevated vein a rapturous encomium. The greatest spectacle in nature is the "leaves and flowers of its silent (spring) concert, its song of March and April, its symphony of May." Michelet emphasizes that birds lived before mankind, but not mankind without birds; that birds protect humans from many insects; that the laborer's bird, in humble dress, is the Lark. If his is an archaic style, nostalgia might call it more than adequate for its subject.

Any modern student of birdsong must be dazzled by the ever-increasing scientific literature of bird vocalizations. Donald Kroodsma and Edward Miller have collected some prominent essays, as had R. A. Hinde a quarter century before. Many of these studies offend the squeamish or the sensitive. For a German study reported in 1960 in *Behavior,* Turkeys were surgically deafened. Although they brooded on the eggs until they hatched, they then killed their young presumably for their lack of the familiar *peep*ing bond (Schleidt 1960, 254). In a more famous instance of surgery to isolate the sources of birdsong, Fernando Nottebohm sectioned the tracheosyringeal branch of the hypoglossal nerve of the male canary. This is no minor surgery; the birds are nearly muted when it's the left syringeal nerve. To see whether birds harbor inborn "templates" of songs, or if they learn their songs by hearing them, a scientist may remove the cochlea from several birds, rendering them deaf. Such laboratory procedures reveal the anatomy and biology of birdsong, for instance, the role of the medulla. Nottebohm includes a photograph of a cross-section of a male canary medulla showing the commissura infima, which joins right and left halves of the medulla at the level of the tracheosyringealis part of the hypoglossal nucleus. If you can say that last phrase three times fast, you can imitate any birdsong in this book.

A 2001 article in the *New Yorker* portrayed Nottebohm's and Ofer Tchernichovski's researches on mapping the changes in Canary brains as they (the birds, not Nottebohm) learn to sing (Specter 2001). Tchernichovski designed and built a sound system into a plastic model bird. He and his researchers then placed month-old Canaries in a sound-studio cage with this plastic daddy-oh wired for song. The young birds responded immediately to the songs of the plastic bird, which can be programmed by the researchers. If a researcher whistled Gershwin, the plastic bird could immediately imitate it and "teach" it to the youngsters. However, according to Michael Specter, the author of the article, "Gershwin is a bit too complex for a songbird to master."

Perhaps. I have heard passerine songs more complex than Gershwin melodies, however—those of Orioles, Red-winged Blackbirds, Starlings, Wood Thrushes, Mockingbirds, and certain Warblers. More discouraging for our purpose is the inducement offered these birds for learning. Tchernichovski says: "Now, if we want to say a certain note was learned at a certain instant, we can take the bird and sacrifice it the second we see him learn that note" (Specter 2001, 53). Somehow, except for suicidal Canaries, being "sacrificed" doesn't seem the best inducement for learning.

Despite great advances in knowledge of avian neurology, and similar advances in bird behavior studies, we still can only guess at how birds use their songs or language. Even to call it a "song" is to misrepresent what may sometimes be more like spitting than singing. A human behavior comparable to territorial birdsong may be the Western film cliché of the gunfighter hitching up his gun belt. As for the usual scientific experiments, one involves playing birdsongs back to the birds that made them. If the birds are at all like adolescent human would-be rock stars, they enjoy hearing themselves recorded. But often the experimenters are more rigorous and unforgiving. It's a wonder they don't drive the birds wacko. Such scientific articles often have pictures of the recorded bird—say, a Starling—standing atop the irritant speaker box, giving it hell.

The Nineteenth-Century capture model of ornithology may have grown into such modern scientific study where we murder to dissect. It also resulted in federal laws, the first being the Migratory Bird Treaty Act of 1918, part of international action to halt the decline in migratory bird population:

> The primary threat to migratory bird populations when the Act was passed was unrestrained shooting for commerce and sport. Consequently, the main thrust of the Act and the treaty [with Canada] is to regulate "taking" of migratory birds, especially hunting. (Bean and Rowland 1997, 64)

In one case resulting from Congress's act, *Missouri vs. Holland* (1920), comes the stark prediction, "But for the treaty and the statute there might soon be no birds for any powers to deal with" (Bean and Rowland 1997, 63). Later federal laws (1940 and 1992) have protected the Eagle, with exceptions for Native American religious practices (465).

Wood Thrush

Avitherapy

You can talk to your friends, or if you have no friends enthralled by details of your latest body fluid samples and intestinal problems, you can talk to a physician—who is usually too busy to listen after a couple of minutes. If your problems are more pressing, you can talk to a psychiatrist.

Or you can talk to a bird.

Birds charge less—considerably less. And their conversation is more beautiful, even if your shrink is operatic. Birds won't shower you with jargon based on bad translations. (Freud never used the term *ego,* for example, which came in with the 1933 English translation.) Avian language may well be the language of the *psyche,* of the soul and mind.

A good reader of my book must have inferred by now that birdtalkers have found the cheapest therapy. I do not deny it. In fact, in discussions with bird enthusiasts, I have found that their interest often grew fastest during a period of personal crisis. Some wonderful ornithological works also acknowledge the sanative effect of birds, especially birdsong. Quite often books on birds are departures, the culmination of academic careers in other disciplines, such as the emeritus psychologist Harold Burtt's *The Psychology of Birds.*

Consider the Nineteenth-Century French popular historian Jules Michelet, cited in the previous chapter. For forty years (from 1833 to 1872) he published twenty-seven volumes in the history of France from the Middle Ages to 1815. Born the son of a printer imprisoned for debt in 1808 when young Michelet was ten, Jules thrived as a student and eventually became a history teacher at École normale supérieur, and later at the Collège de France. Suspended at the revolution of 1848, reinstated under the Second Republic in March, suspended anew two years later, his permanent dismissal followed a year later when he refused to take the loyalty oath to the Empire (Guthrie 1942, 229). After such treatment at the hands of his fellow beings, and after lifelong study of their even worse treatment of each other—called "history"—is it any wonder Michelet found solace in the birds?

He says as much in his preface to *L'Oiseau,* "How the Author was Led to the Study of Nature." He feels he owes his faithful public, who have not abandoned him over such a long time, the private circumstances that led him to "hatching" this book—"if it is a book." He implies it may be an egg, one overseen by his collaborative friends and especially his other collaborators: *les hirondelles familières* (nesting Swallow), *le rouge-gorge domestiques* (Robin), *parfois le rossignol* (Nightingale).

> The heavy times, the life, the work, violent changes of our age, the exile from the intelligentsia among whom we lived, nothing has replaced. The hard work of history finds relaxation in teaching, which complements it. The interruption of that left only silence. To whom look for relaxation and moral refreshment, if not to nature? (Michelet 1876, 4)

The Eighteenth Century alone contains a "thousand years of

combat" from which Jacques-Henri Bernadin de Saint-Pierre recovers with the touching words of Ramond: "So many irreparable losses wept for on the breast of nature!" As for Michelet, he looks for other consolation than solitary tears and sayings that sweeten the wounded heart, "We seek a stimulant cordial to keep moving forward, a restorative sip from an inexhaustible source, a new force, one with wings!" (5). And Jules Michelet is not alone.

Others have found nonhuman succor in nature. In literature, Jonathan Swift's Gulliver returns home after his fourth voyage to the land of the philosopher-horses, the Houyhnhnms. Gulliver precedes us in cross-species conversation. After months of talking to the high-minded horses, he grows disgusted with human behavior and human language. The Houyhnhnms, you recall, had no word for lying. Rational creatures, they could see no reason for language except to express truth.

Probably the most famous turning to birds in the midst of personal crisis is John Keats's "Ode to a Nightingale." Despairing over the loss of his younger brother and over the wasted, indifferent beauty of the spring morning that surrounds him, Keats feels drugged as if by hemlock or opium. At first hearing the Nightingale he feels he wants to fly like that creature, but as he listens he finds himself envious of the bird as singer. The bird has a kind of impersonal immortality in his song, more or less the same as it has been since the days of the biblical Ruth. The bird has achieved what Keats looks for, a song that lives on through "Negative Capability," where will and personality are replaced by art. Keats's depression has been treated by avitherapy.

Moving from literature to real life, take the example of a returning student of mine, the mother of five. Married to a dependable husband with artistic training, she and her whole family once became stranded in a snowstorm in an isolated cabin in Canada. An accompanying friend who did not take off her hooded jacket for three days, even in the cabin, asked at one point, "Why are we doing this?" My student's husband began shaking—at first, apparently from the cold. But upon seeking the nearest physician, they found out he was suffering a nervous breakdown. The physician recommended visiting family—in Arizona? or Hawaii? They did. (Shall we call it the Italian cure?) Ever afterward, my student recalled the White-throated Sparrows on the bird tray out the window at the motel where the family recovered from the freezing cabin. And then she remembered the beautiful songs of the stunning Cardinals out the window of their cousin's cozy ranch house. Simple pleasures. And health restored through birdtalk.

Birds speak theatrically, as do we who respond to them. Humans gather closer together to communicate. Usually we speak at less than the distance of our own height. Birds, on the other hand, mount a scaffold or climb a bell tower. The highest tree. To humans, they all seem to be would-be politicians at national conventions. But that is not entirely true; even their common conversation takes place over substantial distances. So birdtalkers self-dramatize. Such self-projection is a common ploy in other forms of therapy. Clear the air. Wave your arms (or wings). It's good for you.

You have a choice of avitherapists, just as you do psychologists and psychotherapists. I began with that most cheering and practical of advisers, the American Robin. He's a burbler and a bubbler, rather impure, not flutelike. Insistent and repetitive but varied, his overall advice is cheery and direct. Do this. You can do it. Get at it. Don't delay. I think of him as a saxophonist, but not the lilting Paul Desmond or Stan Getz. Robins are closer to the rock-and-roll saxophonist.

If you suffer from obsessive compulsions, you have two choices. You can birdtalk to a truly obsessive species, such as the Titmouse. (See the philosopher Hartshorne on many species that have a low monotony threshold!) Or you may choose one of the great riffing species who never seem to repeat. The Mockingbird is an obvious choice here; others include Redwinged Blackbirds and Warblers.

For bipolar problems, you can choose a truly conflicted bird such as the Woodcock, who on the ground sounds like a large buzzing insect and in the air like a hundred kisses. Or you might be better off with a steady-Eddy sort, the Whitethroated Sparrow.

Depressives have a wide range of options. Almost any birdsong takes one outside oneself into the world of a creature so small, yet so present. In America, depressives are more fortunate than John Keats, for we have the most beautiful of singers, the Wood Thrush. Its pentatonic tunes, which modulate surprisingly from key to key, suggest melancholy, but they also suggest almost total bliss. Less blissful, the Mourning Dove sounds dignified, almost funereal, although he may be having the time of his life. Depressives who talk with Mourning Doves may find a perfect "objective correlative" for their interior feeling.

Robin

Literary Birds

Some of the worst lines in English poetry were inspired by birds. Consider Shelley's infamous "Bird thou never wert" and Emily Dickinson's "But were I Cuckoo born" (when my students usually think she was born cuckoo). Yet even this unfortunate line, a line that colloquial anachronism has killed—as it has W. B. Yeats's favorite adjective "gay"—even Dickinson's fatal line appears in one of her half-dozen fine poems on Robins. She also has famous poems on the Hummingbird and the golden Baltimore Oriole ("One of the ones that Midas Touched"). Birds have elicited some of the best lines, too—in Keats, in Shakespeare, in Catullus, in Virgil, and in Yeats, to name a few. Many of these bird references are exalted, but we start with the humble.

Dickinson's Robin poems rank as definitive. Her Robin is a metaphor for the New England farmer, even the working classes, and a spokesman for conventional values. I start with her later verse, in reverse order of composition:

> *The Robin is a Gabriel*
> *In humble circumstances—*
> *His Dress denotes him socially,*
> *Of Transport's Working Classes—*
> *He has the punctuality*
> *Of the New England Farmer—*

The same oblique integrity,
A Vista vastly warmer—
(DICKINSON 1960, NO. 1483)

We can hear Dickinson's own Nineteenth-Century New England dialect in this last line of the first stanza (or first two stanzas in ballad form). She would have said, "a vistah vahstly wahmah." This I hypothesize from the early pronunciation of the place-name Amherst. A teacher of mine who came to Amherst in 1928, Theodore Baird, had it from a colleague who had arrived in the waning years of the Nineteenth Century, when Calvin Coolidge was the college valedictorian. Amherst was then pronounced *Amst.*

More than a dozen years before she wrote this almost Whitmanesque idea celebrating Gabriel Robin's social class, Dickinson wrote one epigrammatic stanza:

The Robin for the Crumb
Returns no syllable
But long records the Lady's name
In silver Chronicle.
(NO. 864)

Earlier that year—a couple of years after her annus mirabilis of 1862—she interprets the Robin's silence and renders its song bursts precisely:

The Robin is the One
That interrupt the Morn
With hurried—few—express Reports
When March is scarcely on—
(NO. 828)

That's it. Dickinson here lays down the best description of the American Robin's song, a burbling, impure but emphatic series. Most people over a certain age—six years in the suburbs, ten in the cities?—can identify a Robin when they see one, but how many know it by its song? Far fewer. One in a hundred, a thousand. Yet its song is unmistakable. And it sings during light rain, unlike many other birds. This pluvial singing St. Francis would approve, its grace before meals.

"The Robin is the One" is what Yeats calls "passionate syntax," a poetic thwarting of the limits of English grammar: here, the poetic singer uses "one" as a collective pronoun. Neat trick. As a collective pronoun, it takes a plural verb, "the One / That interrupt."

It's grammatically impossible, but it works. A Robin is both one—and many. More of this when we look at Keats's "Ode to a Nightingale." Pablo Neruda aims for a similar paradox in his opening poem to *Art of Birds*. "Migration" begins with: "All day, column after column, / a squadron of feathers," and later observes "nothing but orderly flight" (Neruda 1985, 15–17). Perceiving the many as one, Neruda sees

> *Above the water, in the sky,*
> *the innumerable bird flies on,*
> *the vessel is one,*
> *the transparent ship*
> *builds unity with so many wings,*

Though I do not know many of the birds he writes of, his volume strikes me as the definitive austral bird literature.

Later in Dickinson's Robin song poem she interprets what Robins say not by their speech but by their silence.

> *The Robin is the One*
> *That speechless from her Nest*
> *Submit that Home—and Certainty*
> *And Sanctity, are best.*
>
> (NO. 828)

Dickinson interprets the Robin's values as what we call "home, motherhood, and apple pie." Of course, the female Robin nests for only ten days, although she—and her mate—feed and teach offspring to forage for another couple of weeks. According to one source, the male Robin "cares for the fledged first brood while the female incubates the second batch" (Ehrlich et al. 1988, 462).

Dickinson's earliest Robin poem, with which we began, is fine, despite its unfortunate line. About comparative standards, Dickinson is locative:

> *The Robin's my Criterion for Tune—*
> *Because I grow—where Robins do—*
> *But, were I Cuckoo born—*
> *I'd swear by him—*
> *The ode familiar—rules the Noon—*
>
> (NO. 285)

Later in the poem, she says "were I Britain born," she'd prefer nuts to daisies, disbelieve in snow, because even "The Queen, discerns like me— / Provincially."

One who *was* "Britain born," John Keats, wrote the most famous bird poem in all Western literature, "Ode to a Nightingale." The title suggests the kinds of encounter that saint Giovanni di Bernadone and I have had in such differing ways, that is, an exchange between species. But Keats is no St. Francis, with his homiletic injunction to avian gratitude; still less is he like yours truly, simply echoing and practicing bird language. Rather, Keats is a Romantic poet, and consequently he talks more to himself than to some birdbrain. He meditates on his heartache, his almost drugged numbness, his world weariness, which is punctuated and interrupted by the night song of the Nightingale. For me, a problem in translation arises here because the closest American birdsong may be the Mockingbird's. (This raised a problem for Columbus, too, as we shall see.) At any rate, the song or chatter of the Nightingale catapults Keats out of his local depression. He wants to become like the bird, first seizing on the quality of flight: "Away! away! for I will fly to thee." Realizing that such a wish to fly suggests either dreams or drunkenness, Keats denies it, "Not charioted by Bacchus and his pards," the most literary denial of inebriation in all literature. Picture, for instance, an officer stopping Keats and testing him for drink; Keats protests, "Officer, I am not charioted by Bacchus." In his next line he immediately adds he will fly "on the viewless wings of poetry," but then drops this rather grand claim. In fact, he discovers somewhere in midpoem the real quality of the bird he admires and emulates—its song. That is what he wants to become, the singer, the immortal singer:

> *Thou wast not born for death, immortal Bird!*
> *No hungry generations tread thee down;*
> *The voice I hear this passing night was heard*
> *In ancient days by emperor and clown:*
> *Perhaps the self-same song that found a path*
> *Through the sad heart of Ruth, when, sick for home*
> *She stood in tears amid the alien corn;*
> (FERGUSON 1997, 511)

"No hungry generations tread thee down": Does this mean, quite simply, that Nightingales are not game birds, nor are they much subject to predation? I don't know the history of hunting well enough, but this seems to be Keats's point. Perhaps even more, the song cannot be eliminated—except, as we know too well, and as did not occur to Keats, by extinction.

In the Nightingale ode, Keats knows immediately he envies the bird. He thinks at first he envies its capacity for flight. He tries flight as a metaphor, only to abandon it midpoem when he realizes that the claim that he is flying sounds drunken. This reminds us of how young this writer is. Keats drops his desire to fly for his real desire, that of any poet-singer, that his songs continue. Like the Nightingale's. Curiously, the bird's immortality is couched in its anonymity, its indistinguishability from other Nightingales. The one and the many. "The voice I hear this passing night was heard / In ancient days," Dickinson put it, "The Robin is the One." We can say that both Keats and Dickinson are specists. They can't tell one Nightingale, one Robin, from another. But this artful anonymity—to use Keats's phrase, this "Negative Capability"— assures the continuance of the song long after the singer departs.

Of course, Keats was somewhat wrong on "the voice I hear this passing night," which I can guarantee was not identical to any other heard on that night or "in ancient days by emperor and clown." As I have averred repeatedly, with scientific and experiential substantiation, each bird of each oscine species can be distinguished by its song, but usually the distinctions elude nonpasserines like us. Still, Keats's overall point is well taken, that song immortality requires a kind of anonymity. Neapolitan songs like *O sole mio* are sung today regardless of who specifically wrote and recorded them. And the poetry of any given language depends finally on the fortunes of that language, the ethnolinguistic "species" for however many centuries it may endure. For the student of Latin and Hebrew and Hindi, of Catullus and Solomon, "The voice I hear this passing night was heard / In ancient days."

Christopher Columbus the Genovan in Spanish employ made a classic mistake identifying a Nightingale. Leonardo Olschki has pointed out how Columbus functions by bringing what he knows—especially, his language for familiar objects—to what he does not know (Olschki 1937). We all do this. In his case, Columbus records in his *Diario* hearing an *ussignolo* on Hispaniola. Of course, there are no Nightingales in the New World. What did he hear? My guess is some variety of Mockingbird, especially since Ehrlich et al. tell us, "unmated males sing at night in spring" (1998, 468).

The Nightingale is the most literary of birds, like springtime, a literary idea. Calendar spring includes some horrible weather, a high incidence of suicide for males in March, and many disappointments, but the literary convention of spring incorporates none of these. Similarly, the Nightingale is a noisy chatterer, but its literary associations suggest plaintive and unexpected melodies. In fact, the American Wood Thrush delivers what the Nightingale only symbolizes. The Wood Thrush sings at dusk in tall trees, lingering, resonant sequences, variations on a pentatonic scale. If you play all the black keys on a piano, that's a pentatonic scale. But piano cannot render the Thrush's melodic suspension. The notes seem to float on the dark shadows under the tall trees, like a flute.

Often it sings two three-note series, the first two notes a descending minor third (to the tonic), the second and third a rising diminished seventh, this last note lingering. That's the first three notes. The second three start from the same note or one note higher in the pentatonic scale, rises one tone, and falls either a sixth or a fifth. And Thrushes often shift the base tonality, say from the key of E-flat—what sounds so, to one trained in tonal music—they may resolve to an F. At least, so I hear their deeply intriguing song.

Before I move on to Shakespeare's and Catullus's birds, I linger over Dickinson. Her Oriole, what Thoreau calls a Golden Robin, elicits *pennis*-envy. (God'll get me for that, since Dickinson's poem is a masterwork, a toccata or virtuoso arpeggio. And after all, Keats's poem is about *pennis*-envy, too. *Pennis,* the Latin word for "wings," is pronounced the way we say its cognate *pen* plus *ees.* The pun works better on paper.) But surely *envy* is not too strong a word for her second line, her comparison of the Oriole's plumage to human plumage and her own by implication.

> *One of the ones that Midas touched*
> *Who failed to touch us all*
> *Was that confiding Prodigal*
> *The reeling Oriole—*
>
> *So drunk he disavows it*
> *With badinage divine—*
> *So dazzling we mistake him*
> *For an alighting Mine—*
>
> (NO. 1466)

This is such a fine, extravagant poem that it's hard not to keep quoting, especially because the next stanza describes the Oriole's song, "A Pleader—a Dissembler— / An Epicure—A Thief— / Betimes an Oratorio— / An Ecstasy in Chief—." What I have been analyzing as the complex music of the Oriole, Dickinson calls "an Oratorio" and "badinage divine." Who could disagree?

I think her "reeling Oriole . . . so drunk" comes from their feeding habits around, say, locust tree blossoms. I'm not sure whether they feed on the blossoms or on the bugs that feed on the blossoms. Trusty Ehrlich et al. say "spiders, some buds in spring," but they grant nectar-sipping to the Hooded Oriole (1988, 624). In short, Orioles do move around as they feed. Upside down, sideways, and so on. Of course, many birds do this—Nuthatches, famously—but perhaps the Orioles' brilliant coloration makes their motion more obvious.

Dickinson calls them "the Meteor of Birds, / Departing like a Pageant / Of Ballads and of Bards." My other Amherst informant, Theodore Baird, once wrote me that those other brilliant birds, flocking Evening Grosbeaks, at his feeder seemed like businessmen, well dressed, aggressive. So whether birds seem sober or drunk depends upon how they must feed and whether they're migrating, or Zugunruhe.

Finally, Dickinson's spectacular close. Like a Neapolitan author imbued in Greek and Roman myths, calling one coastal town for Aeneas's steersman Palinurus, another for his trumpeter Misenus, Dickinson becomes playfully mythy. Here's her Oriole.

> *I never thought that Jason sought*
> *For any golden Fleece*
> *But then I am a rural man*
> *With thoughts that make for Peace—*
>
> *But if there were a Jason,*
> *Tradition bear with me*
> *Behold his lost Aggrandizement*
> *Upon the Apple Tree—*
>
> (NO. 1466)

Wow. Talk about a pageant. Of course, the poet takes a meta-mythic, skeptical attitude that is reminiscent of Seventeenth-Century English poets like Herrick and Marvell: "Fauns and Fayries do the Meadows till / More by their presence than their skill" (McDonald 43). Before we leave Dickinson's delicious avian evocation—admittedly, more visual than aural—we should note the poet's gender bending in "I am a rural man." Writing in the Nineteenth Century, Dickinson chose, like George Eliot, the male persona, the conventional first person. *Chose* may be the wrong verb since it was foisted on the writer who planned to publish. Dickinson rejects publication in one famous poem, "Publication is the Auction / Of the Mind of Man" (no. 709). But then, she uses the published author's male persona.

Before leaving this great poet and fine ornithologist, I must note her Hummingbird. It is couched like a riddle.

> *A Route of Evanescence*
> *With a revolving Wheel—*
> *A Resonance of Emerald—*
> *A Rush of Cochineal—*
>
> (NO. 1463)

Cochineal is a bright-red dye, so this exactly describes the only eastern species of Hummingbird, the Ruby-Throated, where only the male has a red throat. Both sexes have the bright metallic green above, "A Resonance of Emerald."

> *And every Blossom on the Bush*
> *Adjusts its tumbled Head—*
> *The mail from Tunis, probably,*
> *An easy Morning's Ride—*
>
> (NO. 1463)

Every blossom of, say, the orange trumpet-vine, nods as the fan-jet Hummer goes by. It appears and leaves almost magically, it "evanesces." One witnesses amazing stationary hovering and then speed, like a rocket from Tunis.

Dickinson reenacts for us the experience of her seeing birds, like her other Robin poem.

> *A Bird came down the Walk—*
> *He did not know I saw—*
> *He bit an Angleworm in halves*
> *And ate the fellow, raw.*
>
> (NO. 328)

After this brutal, thoroughly natural, candid event ("He did not know"), the Robin shows his civil grace by drinking "a Dew" from a "convenient Grass" (punning on "glass") and hopping aside to let a Beetle—otherwise a candidate for consumption?—pass. Dickinson observes the Bird's "frightened Beads" of eyes, how it seemed "like one in Danger." A brilliant observer of avian behavior, this poet recognizes the menacing world birds populate.

Shakespeare, on the other hand, appears to cite lore about birds as much as his own immediate impressions. For instance, he knows about the fascinating phenomenon we in America call the "Cowbird host." Cowbirds (*Molothrus ater*) lay their eggs in any nest, and the female experiences no ovarian shrinkage after laying, so they have been termed "passerine chickens" (Ehrlich et al. 1988, 619). One species of Warbler, Kirtland's, has gone nearly extinct because it can't distinguish Cowbird eggs from its own, and the cowbird young, being bigger and more demanding, get the "Cowbird's share" of its dutiful though ignorant foster parents' harvest. Cowbird young also occasionally bite the parental feeder by mistake, sometimes fatally.

Cuckoos (*Cuculus canorus*) are another brood parasite, and they have been photographed pushing the eggs of rightful offspring out of the nest. Shakespeare mentions the Kirtland's Warbler phenomenon in *King Lear*. As Goneril complains about her father's "insolent retinue" and about the Fool's freedom of speech ("this your all-licensed fool"), the Fool sings snatches of a song about Cuckoo hosts.

> The hedge-sparrow fed the cuckoo so long
> That it's had it head bit off by it young.
>
> (1.4.198)

No sentimentalist, Shakespeare does cite lore and convention. His Sonnet 29 features the Lark as an image of gratitude to the creator, with or without Franciscan instruction. On thinking of his friend, the speaker recovers from his blues, from his solitary depression ("I all alone beweep my outcast state"), from his feeling "in disgrace with . . . men's eyes." His mood, his "state" changes, "like to the lark at break of day arising / From sullen earth, sings hymns at heaven's gate."

But it would be hard to find better bird writing in Shakespeare than on those architects and domestic builders, Purple Martins. Perhaps it is fitting that the owner of New Place, the second largest house in Stratford, with its fourteen fireplaces, would notice avian housing. Toward the beginning of *MacBeth,* King Duncan arrives at the MacBeths' castle. He comments on the beauty of the place, its "pleasant seat," the nimble and sweet air. Banquo observes—high in the rafters of the Globe Theater:

This guest of summer,
The temple-haunting martlet, does approve
By his loved mansionry that the heavens' breath
Smells wooingly here. No jutty, frieze,
Buttress, nor coign of vantage but this bird
Hath made his pendant bed and procreant cradle;
Where they most breed and haunt I have observed
The air is delicate.

(1.6.3–10)

I must confess I have never seen Purple Martins in "pendant beds," although I know the birds from childhood. They competed with my grandfather's Barn Swallows. The Swallows nested under the barn roof; the Martins inhabited my grandfather's large, two-story birdhouse, on a ten-foot pole. So I never saw their hanging nests, but I have seen abandoned Oriole nests overhanging roadways in winter, as did Thoreau one December:

> As we passed under the elm beyond George Heywood's, I looked up and saw a fiery hangbird's nest dangling over the road. What a reminiscence of summer, a fiery hangbird's nest dangling from an elm over the road when perhaps the thermometer is down to -20 (?), and the traveller goes beating his arms beneath it! It is hard to recall the strain of that bird then. (Allen 1993, 288)

Ah, the song. Thoreau brings us back to our theme.

As for Purple Martins, Shakespeare also mentions them in *The Merchant of Venice,* where Aragon cites the Martlet as an image of common judgment,

Which pries not to th' interior but, like the martlet,
Builds in the weather on the outward wall
Even in the force and road of casualty.

(2.9.26–29)

These hanging nests, marvels of avian architecture, suggest the parrots nesting in Chicago, Jackson Park, and even an alley off Fifty-fourth Street, in Saul Bellow's *Ravelstein:*

Like over-stretched nylon stockings, those nesting tenements where eggs were hatched drooped as much as thirty feet. (2000, 170)

Suspension nests, like suspension bridges, are marvels. Writers from Shakespeare to Thoreau to Bellow remark them in evocative language. English, too.

Other literatures evoke famous birds. The Romans had Catullus's well-known poem to his girlfriend's Sparrow, "*Passer, deliciae meae puellae*" (Cornish 1966, 2). It's about a pet, not a wild bird. For the Romans, birds were either food or pets. Or en masse, natural signs to be interpreted by a member of the college of auspices (like Cicero), if not a haruspex. Maybe those guys were better on bird flight than on entrails. Often they'd open up some beast and not find its heart. A very bad sign, they thought—though they didn't realize it was a bad sign for their training—the past, not the future. In Catullus's poem, he wishes he had a pet to relieve his loneliness, although he implies envy of the pet's intimacy. A *précieux* poet in Seventeenth-Century France would have stated what Catullus implies, his wish to take the liberties a pet takes, nibbling index fingertip (*cui primum digitum dare appetenti*), pecking or love bites, lying in her lap.

Return to the Robin we began with, but now the smaller English Robin much celebrated in poetry. The Seventeenth-Century preacher and wit Robert Herrick wrote: "Upon Mrs. Eliz: Wheeler, under the name of Amarillis." In this twelve-line epigram a "Robin-Redbrest . . . brought leaves and mosse to cover her" until "Eliz" opened her eyes—which in the old science allows light to come out: "The lid began to let out day" (Herrick 1960, 49). A century later, William Cowper delights in this bird in midwinter:

> *The redbreast warbles still, but is content*
> *With slender notes and more than half suppressed:*
> *Pleased with his solitude, and flitting light,*
> *From spray to spray, where'er he rests he shakes*
> *From many a twig the pendent drops of ice*
> (LACK 1965, 32)

Again, probably the most famous English Robin reference occurs in Keats, this time in his "Ode to Autumn." In answer to the question, "Where are the songs of spring? Ay, where are they?" Keats writes some of the most lovely aural lines in English verse:

> *Then in a wailful choir the small gnats mourn*
> *Among the river sallows, borne aloft*
> *Or sinking as the light wind lives or dies;*
> *And full-grown lambs loud bleat from hilly bourn;*
> *Hedge-crickets sing; and now with treble soft*
> *The redbreast whistles from a garden-croft;*
> *And gathering swallows twitter in the skies.*
> (FERGUSON 1997, 514)

What a closing! With the previous sounds of the cider-press oozings, the late bees, and now the lambs, the crickets, Robins, and Swallows, Keats challenges the great literary dominance of spring. Spring is, after all, a literary idea, not the "real" spring of mud and flood, chill and roof leaks. In this ode, Keats chalks up one for Team Autumn.

Finally, Herrick's predecessor Edmund Spenser includes the English Robin ("Ruddock") in his exotic catalog of summer birds.

> *The merry Larke hir mattins sings aloft;*
> *The Thrush replyes; the Mavis descant playes;*
> *The Ouzell shrills; the Ruddock warbles soft;*
> *So goodly all agree, with sweet consent,*
> *To this dayes merriment.*
>
> (LACK 1965, 32)

A fine example of Elizabethan plenitude in writing, Spenser repeats himself since both the Mavis and the Ouzell are Thrushes, as is, arguably, the Ruddock. But who demurs with taxonomical correctness?

The English Robin, much smaller than the American, certainly compares in aggressiveness. In *The Descent of Man and Selection in Relation to Sex* (1871), Charles Darwin—who may be called a literary author in that he calls nature a "she"— remarks on the bird's fatal aggression in captivity.

> Mr. Weir was also obliged to turn out a robin, as it fiercely attacked all the birds in his aviary with any red in their plumage, but no other kinds; it actually killed a red-breasted crossbill, and nearly killed a goldfinch. (Baird 1946)

Even in the wild the English Robin displays similar behavior, as David Lack writes after quoting Darwin.

> That intruders are recognized is evident enough, since in nature the owning robin promptly drives from its territory all other robins which try to enter, the bird's own mate only excepted. On the other hand, birds of other kinds are not usually attacked. (Lack 1965, 165)

I seem to have crept into my next chapter, on foreign birds, already. That's the nature of literature; it takes one beyond one's own experience.

Before moving on, probably the second-most-famous bird in English literature, after Keats's Nightingale, is the "dapple dawn-drawn Falcon" of the Nineteenth-Century English Catholic priest Gerard Manley Hopkins. (Recall that Robert Herrick, too, was a priest, an Anglican.) Of course, Hopkins's capitalized "Falcon" is not our mere species name, it is a Christological image, certified by the title "The Windhover," with the epigraph "To Christ our Lord":

> *I caught this morning morning's minion, kingdom*
> *of daylight's dauphin,*
> *dapple-dawn-drawn Falcon, in his riding*
> *Of the rolling level underneath him steady*
> *air, and striding*
> *High there, how he rung upon the rein of a*
> *wimpling wing*
> *In his ecstasy!*
>
> (FERGUSON 1997, 662)

Hopkins apostrophizes "the achieve of, the mastery of the thing!"—the thing being virtuosity of flight, which the poet takes as an image of religious ecstasy, and perhaps of Christ's rising from death. This grand poem should be quoted in its entirety, but it describes bird *flight,* not song. The Falcon doesn't speak. If he did, I suppose it would be heresy. And, of course, you could argue it's not about a bird at all.

Similarly, Andrew Marvell's justly famous "The Garden" includes these birdy lines comparing the soul to a bird.

> *Casting the Bodies vest aside,*
> *My Soul into the boughs does glide:*
> *There, like a Bird, it sits and sings,*
> *Then whets, and combs its silver wings;*
> *And, till prepar'd for longer flight,*
> *Waves in its plumes the various light.*
>   (MacDonald 1966, 53)

Marvell's lines seem even less about a bird than Hopkins's. Both poets are religious or philosophical and theological. Marvell's birdlike soul apparently derives from Plato and the Neoplatonics. In his next stanza, Marvell launches into a witty argument about prelapsarian man in his "happy garden-state / While man there walked without a mate." He ends this stanza with the puzzling ironic paradox, "Two paradises 'twere in one / To live in paradise alone." I am tempted to call this anti-Romantic, but Marvell seems to recommend solitude, whereby he anticipates the Romantic poets—and of course prepares the way for the Romantics' rediscovery of his poetry.

Asian cultures offer marvelous birds in art and poetry,

my single example coming from China, Li Bo's "Waking from Drunkeness on a Spring Day," where the poet feels life is a dream—with Spanish playwright Lope de Vega and many Shakespeare characters (Li Bo 1988). In Li Bo's case, the speaker uses it for an excuse to drink:

> *Lying helpless at the porch in front of my door*
> *When I woke up, I blinked at the garden-lawn;*
> *A lonely bird was singing amid the flowers.*
> *I asked myself, had the day been wet or fine?*
> *The Spring wind was telling the mango-bird.*
> *Moved by its song I soon began to sigh,*
> *And as wine was there I filled my own cup.*
> *Wildly singing I waited for the moon to rise;*
> *When my song was over, all my senses had gone.*

Here is an epigrammatic version of Keats's "Ode to a Nightingale." Keats longs for a "draught of vintage"; in the English way, his pleasure is deferred, giving rise to the great literature of the isle. Had Keats drunk more, he may have written less, like Li Bo. The metaphor of birdsong as poet's song is identical, although Keats first wants to be like the bird in flight. Drunk, it's a good thing Li Bo doesn't explore that bird quality.

In short, poets have always observed and described birds and found in them an image of exaltation or skyey message. Catullus's Sparrow-pet, Spenser's Ruddock, Shakespeare's Lark and Martlet, Herrick's Robin, Keats's Nightingale and gathering Swallows, Dickinson's Oriole, Robin, and Hummingbird, and Hopkins's Falcon all form a vast catalog. We may call it an aviary, attended and nursed by the literary imagination.

Titmouse

North Atlantic Birds

*B*ird is a universal language, a lingua franca, like Esperanto. At the risk of sounding like an ad in the *Times* travel guide, after a few days' practice you really can talk to natives in any country—France, Italy, Brazil, Japan, China. Of course, you'll be talking to native birds. You're unlikely even in suspicious nations to be breaking any local spy ordinances.

To start close to home, offshore: one midnight watch crewing on a sail from Jacksonville to Massachusetts, I heard our mast ringing. Some lashing or fitting made a *ping* on the alloy. Behind me over our wake something was echoing that ring. It was following us along in the moonlight, a fairly small bird, I think a Petrel, probably Wilson's. This was in late May. Was it enamored of our boat's mast? Talk about impossible relationships. During the day we saw them all the time, Mother Cary's Chickens. Wonderful fliers, they breed in the Antarctic, as do some Terns.

A couple of years ago I was lucky enough to spend a month on the south coast of England, in Dorset. The month of May, no less. Mixed with the sounds of lambs bleating and occasional cows mooing, now and again a shotgun popped in the near distance across the vale toward the pre-Roman earthen forts, Coney's and Lambert's Castles. Damp and overcast most mornings, there was always a cheering bird singing his dotted eighths, repeatedly. A Blackbird. English Blackbirds really do sing

melodies, not just the inimitable *scree* and *wree* of our marsh-
land variety. Hence Paul McCartney's song, "Blackbird singing in the dead of night / Take your broken wings and learn to
fly / All your life / You were only waiting for / This moment to
arise." That Blackbird's song was imitable, though varied. Here
are my notes—well, really his notes, though at least one of his is
a microtone, the second accented note, between A and G-sharp.

His song swings. In fact, I composed a jazz piece on the basis of
this bird, as I did another based on the pentatonic, bluesy song
of the American Wood Thrush.

   I prefer the English Blackbird because I can imitate it, while
the multivocal Red-winged Blackbird makes dozens, perhaps
hundreds of inimitable vocalizations. Les Beletsky classifies the
"alert" calls alone into: *peet, check, chuck, chick, chonk, chink,* and
cheer. What I termed a *scree,* he calls the territorial call note a *conc-
a-ree.* There turns out to be much scientific study of Red-wing
song because of "the relative ease of conducting field studies on
the species" (Beletsky 1996, 84). Red-wing marshes are relatively
confined spaces with high bird populations; hence the birds' social
and vocal behavior can often be surmised. For instance, speakers
broadcasting recorded Red-wing song were aggressively perched
on by male Red-wings and approached even by females.

   Among those who have studied the Red-wing song are
R. W. Simmers (1975), Douglas G. Smith and Fiona A. Reid
(1979), Ken Yasukawa (1979), and Donald Kroodsma and
F. C. James (1994). Beletsky summarizes their findings by
raising issues that apply to all the bird species we mention:
geographic variation, birds' learning new songs into their

second year, and the effect of their gathering at large winter roosts on uniformity of call over their large range (1996, 87).

Returning closer to home, what about the Loon? Here's a bird whose southernmost habitat lies in our northernmost states. Are Loons foreign? Not exactly. Sharing the lands of the moose, they call amid vast silences, at night. Their nocturnal call is at once lonely and very social, for of course, another Loon responds. They call at other times but not while feeding. I saw a pair in late afternoon on the headwaters of the Androscoggin, just after seeing a Great Blue Heron and, at another section, a young moose eating salad. Some sort of water lily. The Loons dove at will; when I called, the male stretched his neck to see but was not fooled. The pair did converge, though. They continued to preen and raise one wing so they looked white-backed, or rub their necks on their backs—neat trick. They have oil glands in their mouths to "paint" their feathers after diving. A few minutes earlier on this August evening, with occasional trucks and vacationers' vans going by, I heard a Loon across a half-mile cove, beyond an island. I returned its call, and it called back. In broad daylight. I have read that they give one of their calls only when separated, a locative "Where are you? I'm here."

In the second week of August, in Rangeley, Maine, after all the roaring car engines die down around 3 a.m., an occasional Loon calls. Again, theirs is a minor third, flutelike, not a whistle. One of the slowest birdcalls, the notes are held, about one per second. Do they wait for the cackle of humans to die down? The motorcycles and big pickups, the last motorboat with running lights on the lake. I don't hear their "laughing Loon" call, hyena-like. Instead, they flute a quarter note, hold it a second, rise a minor third, and slur it a quarter tone down. The next time, it glissandos or slurs a quarter tone up.

Finally, they start the same, but then end on the initial note.

So plaintive, or is it plangent? The Loon's call goes right to my bones. It seems to tell me something of vastness and timelessness, qualities that Keats taught us to associate with birds. I walk to the side of the lake and hear a couple of splashes under the half moon. Looking up at the moon through birders' binoculars I can see Crater Riccioli, named for himself by the man who named all the visible lunar features in 1651, Giovanni Riccioli. Does our Loon call to the *lunar?* If so, Moon-day, Monday (in French, *lundi*) may also be the Loon-day. Most other birds, owls excepted, call to the sun.

In the land of the sun, near Naples, one summer I heard birds chattering in tall conifer trees. This was at the Villa Vergiliana above the amphitheater at Cuma, one of the oldest in Italy, just now being excavated. Earlier I had heard from the masonry inset windows what I took to be Italian Blackbirds, tuneful, like the English ones. Something about all the masonry in Italian towns lends an echoic quality to sounds at a distance—Italian mamas calling their sons' names, for instance in the Santa Lucia district of Naples. Or birdsong through masonry inset windows. What I'm calling Italian Blackbirds sound burbling and bluesy, not pure noted. They are syncopated so:

Blackbirds are not the only birds common to suburban Italy. Count on roosters at dawn.

Back in the New World, up along the St. Lawrence Seaway, one morning I trod a *sentier ornithologique,* a birders' path. An insistent, repetitious bird called from the conifers bordering the riverbank. It was a rising call, just one tone. A Goldfinch, I think. There are many up there. I've seen a half-dozen feeding in a gravel drive two or three miles from the river. After all, it was thistle time, and the Goldfinch so prefer thistle seed that the French call them *chardonneret,* which means "thistler." The Goldfinch speech I'm most familiar with is an insistent in-flight call, falling just one tone.

One French name for a bird species can be read by those of us who do not have French. It's the onomatopoetic moniker Wood Peewee, spelled in French, *pioui.* Say it like Peter Sellers's Inspecteur Clouseau. *Pi,* pronounced "pee," and *oui,* the French word for yes pronounced "wee." To French ears, does this bird sound like a yes-man? Perhaps. The English name sounds like micturition—pee-pee and wee-wee—but I doubt any native English speaker associates the bird with urination. I have conversed with Peewees on South Mountain in the Berkshires. Even in mid-August, like a Titmouse in spring, a Peewee can be "called" by echoing his song. However, they stay in higher trees, verdure virtually the same as Wood Thrush habitat. From his considerable height, the Peewee makes his onomastic sound, a fairly piercing rising glissando. Starting around the E an octave above middle C, the Peewee slurs up just shy of an F-sharp. To my ear, it's not exactly a full tone, although I could be wrong. The more I concentrate on birdsong, the more subtle and unrepresented by human notation it seems—even the modern notation in Gardner Read's books. Lots of variable birdsong quarter tones, for instance.

There surely is a cross-cultural as well as a cross-species aspect to birdsong. Onomatopoeia reveals a culture. Take the American sound for a cat, *meow,* and the British *mew.* Which cat sounds better bred? Which would you prefer in your house? Which would you prefer to stalk the attic for those pesky mice that scratch all night? Or consider the Russian verb for a dog's bark, *lyalyat'*. That's a puzzle for English speakers until you see a Russian film in which Borzois appear. In fact, Borzois on a hunt make exactly that sound, a kind of gulpy, almost gull-like bark—*lya-ee, lya-ee*. Despite the hunt culture they share with the Russians, the English words evoke domesticated dogs, *woof* or *bow-wow*. While I have heard dogs make a suppressed *wuhf*, I have never heard them *bow-wow*. In Italy, dogs *abbaiare*. Although clearly an imitation of dogs, onomatopoetic, it derives from Latin and Greek. According to Battisti and Alessio, the Italian Maritime Alp speech preserved the Latin *baubare* from the Greek *bayzo*.

Horses *whinny*. For American speakers, they do. To the English they *neigh*, to the Russians, they *rjat'*, to the French, *hennir*. But when I heard a carriage horse in Quebec one hot August day, that horse pierced me with the equine condition. A handsome fellow, though no longer young, he was pulling a carriage with an overweight tourist not so different from myself and an overweight driver. His mouth around the bit showed a bit of white foam. His whinny lasted maybe six seconds, *whi-hih-hih-hnih-hny*. It was a cri de coeur from a stout-hearted beast. I thought immediately of Jonathan Swift's marvelous representation of horse language in Book Four of *Gulliver's Travels*. The race of horse philosophers call themselves, as might Eighteenth-Century human philosophers like Leibnitz, *the perfection of nature,* or Houyhnhnm. Their species name is a whinny. They despise the humanoform Yahoos, from which word derives their name for disease, *hnea-yahoo*. Anything bad is named using the suffix *yahoo,* as an ill-constructed house, *ynholmhnmrohlnw yahoo*.

Like Gulliver, I have spent more time learning from animals, very little time in educating birds, though I did try to teach one Oriole the famous theme from Beethoven's Fifth

Symphony. According to the accounts, St. Francis did "preach" to birds, enjoined them to gratitude—which one would have thought they already expressed in many ways. At least, Gerard Manley Hopkins suggests the Windhover flight expresses his ecstasy. Mostly, then, we learn from them. No one has learned more from birdsong than the French composer Olivier Messiaen. He researched birdsong like a field scientist, rising well before dawn in the spring, going to specific spots to hear specific birds. Wood Thrushes—*la grive des bois*. Blackbirds—*le merle noir*. Bluebirds—*le merle bleu*. Owls—*la chouette hulotte*. Larks—*l'alouette lulu, l'alouette calandrelle*. Buzzards—*la buse variable*. And again and again, Robins—*le rouge-gorge*. Half of his six pieces in *Petites esquisses d'oiseaux* (1985) are titled *Le rouge-gorge*. He captures what I cannot imitate. Messiaen's compositions come the closest to that broad spectrum of inimitable birdsong that I have heard from Blackbirds and even Orioles, once you return their call note "dial tones." It's no surprise that Olivier Messiaen's only opera, a massive work, is titled *Saint François d'Assise*. This French composer preceded me in the topic of my book, but his excurses are in music.

Messiaen and his wife made many field recordings of birds all over the world. In two compositions from the fifties, he develops his field notes:

Actually, in *Oiseaux exotiques* [1955–1956], I've linked percussion with woodwinds, brass, xylophone, glockenspiel, and piano, and this percussion obviously plays no part in bird songs; it constitutes a strophic support based on rhythms, verses, and Greek metric feet, and also on Hindu rhythms, all juxtaposed—the strophic form . . . evolves independently of the bird songs, whose freedom is much greater. So there's a blend of strictness and freedom, and, all the same, a certain element of composition in the "bird-song material," since I've randomly placed side by side the birds of China, India, Malaysia, and North and south America, which is to say, birds that never encountered each other. (Messiaen, 1994, 131)

An earlier composition keeps to birdsong, and interestingly, chronobiology, from the predawn awakening into the night.

In *Réviel des oiseaux* [1953], the presentation is much more accurate: there's really nothing but bird songs in it, without any added rhythm or counterpoint, and the birds singing are really found together in nature; it's a completely truthful work. It's about the awakening of birds at the beginning of a spring morning; the cycle goes from midnight to noon: night songs, an awakening at four in the morning, a big tutti of birds cut short by the sunrise, forenoon songs, and the great silence of noon. (131)

Note that Messiaen's exhaustive study of birds led him to adopt their sense of time. Their voluble day really begins predawn, and it seems to end for Messiaen at "the great silence of noon." In a performance by the Boston Symphony Orchestra (March 2, 2002), avian microtones were achieved by dissonant notes on, say, violin and piccolo or the highest octave on the piano in combination with the ethereal celeste. Messiaen builds to the chaotic "big tutti of birds," which does sound like bird chatter at dawn—the mayhem of a spring sunrise. There is a delightful chaos of chirps and percussive rhythms—especially the five-beat rhythm discussed elsewhere. Then a grand pause, utter silence. The piece ends, after the piano's eighty-third to eighty-eighth keys work tirelessly for minutes. It ends with a wood block's simple tick-tock.

Three years after his exotic birds, the composer linked time and color to birdsong in *Chronochromie* (1959–1960). One part of it, the epode, scandalized his audience. Messiaen explains to Claude Samuel:

> It's written for eighteen solo strings—twelve violins, four violas, and two cellos—and it's made up of eighteen bird songs, all birds from France that sing together and who are, in order of entry, our blackbirds; a yellowhammer; the first goldfinch; a chiffchaff; a second goldfinch; a whitethroat; two chaffinches; a nightingale; a fifth and sixth blackbird; a greenfinch; two golden orioles, one of which echoes the other; and, a little later, a linnet and two garden warblers.

These birds enter one after another, in something of a fugue, the entries proceeding on a descending scale; then all the voices go forward imperturbably, one on top of the other, achieving a counterpoint of eighteen parts totally independent of each other for at least ten minutes (Messiaen 1994, 132).

To Claude Samuel's question whether such a thing were possible in nature, Messiaen assures him that especially at daybreak one can hear "extremely complex counterpoints" produced by very different species united only by habitat. This scandalous part of *Chronochromie* gives rise, a quarter century later, to the "Sermon to the Birds" in his opera *Saint François*. Of course, Messiaen's birdtalk is bispecial.

Just as I cannot imitate most Red-wing songs without Messiaen's help (and a full orchestra), humans more generally can imitate only a small proportion of birdsong. Despite this, our songs often invoke those other, feathered vocalists—and not just French classical composers. Our best popular singers and musicians earn nicknames from birds, like the tenor saxophonist Charlie Parker ("Bird"), for whom the famous New York jazz club was named Birdland. Or the intense, beloved Parisian singer Édith Piaf, the "Little Sparrow." The popular songs they play and sing are filled with comparisons and references to birds.

Consider jazz standards such as "Skylark," "Robin's Nest," and Parker's own "Ornithology," century-old songs like "Mockingbird Hill," more recent songs like "When the Swallows Come Back to Capistrano," and rock songs like Paul McCartney's "Blackbird." The American songbook includes many bird references, like the verse from Johnny Mercer's "Too Marvelous for Words":

> *You're just too marvelous*
> *Too marvelous for words*
> *Like glorious, glamorous,*
> *. . . And so I'm borrowing*
> *A love song from the birds*
> *To tell you that you're marvelous,*
> *Too marvelous for words.*

If that one is dictated by the "words-birds" rhyme, it is still a convincing use of it. One does not see it coming from a mile off, as with so many song rhymes: "blue-you," "you-true," "start-heart," and so on.

Occasional popular songs, like the Neapolitan "*'Nu Pasariello Sperzo,*" tell stories of birds—in that case, a displaced Sparrow chased, by lovers, even from the forest it takes refuge in (Paliotti 1992, 92). Many other popular songs refer to singing, such as the laundry-girls' singing in the second verse of the famous "*O sole mio.*"

There's no bird in this song, but the laundry-girl sings in the midst of nature after a storm.

> 'O sole mio
> Sta 'nfronte a te!
> Luceno 'e llastre d' 'a fenesta toia;
> 'Na lavannara canta e se ne vanta
> E pe' tramante torce, spanne e canta,
> Luceno 'e llastre d' 'a fenesta toia.
>
> (PALIOTTI 1992, 126)

Because this is Neapolitan, I shall not give a word-for-word translation but metaphrase instead: "O my sun, keep before us. The glass in your windows is bright with sun, a washerwoman sings and vaunts, meanwhile wringing, clothes-pinning, singing." Our selection begins with the refrain that follows the first verse. Here I *do* translate word for word. "What a beautiful thing it is to see the sun in the serene air after a storm, why it's like a party already started." (*Che bella cosa è 'na jurnata 'e sole, / 'N'aria serena dopo 'na tempesta! / Per l'aria fresca pare gia 'na festa.*)

Jazz songs often conflate birds and women through the colloquial usage of bird, as in Tadd Dameron's "Lady Bird." In British usage, such a reference is entomological, a ladybug. I think not even rock music descends to the vulgar and obscene usage of bird, as in the common gesture of "flipping someone the bird." All bets are off when it comes to rap music—though if biologists are right about the braggadocio of territorial birdtalk, birds are rappers, full of aggression. For the biologists, birdtalk is largely like gang insignia, the marking of territories.

White-throated Sparrow

Winter Birds

Not all birds are taciturn in winter, but don't expect their March chatter. Sometime in late October or November, the migratory species clam up. Even the resident Blue Jays and Titmice and White-throated Sparrows do not respond as the temperature hits freezing, or as their peers are flocking to migrate. This makes sense, after all. Besides weather and especially solar discussions, much birdtalk concerns property rights, their call notes like avian Gold Rush prospecting claims. Surely only a bad-mannered Blue Jay would bring up his spatial needs in the midst of the crowd, the flocking delegates. So it follows that I should have expected precious little response when I called the firm, descending major third that only two days earlier had elicited a Jay flyby over my head. However, today was different. For one thing, there were ten or twelve Blue Jays within two or three small, thirty-foot trees. For another, they perhaps were exhibiting Zugunruhe.

While we may expect reticence, we also receive compensating gifts. Winter birds desperate for food appear at feeders close enough to touch. For birdtalkers, such an encounter is even more thrilling. We meet the little Sparrow with the white throat that we first heard in the deep pine woods of Maine. This January morning it's right here—in fact, a couple of them—on my front porch, its white throat patch outlined by a darker band. I feel as if I'm meeting a famous musician, a Wynton Marsalis or a Nina

Simone. At the same time, I am meeting a correspondent, a pen pal, someone whose language—and moods—I've been learning.

I find this meeting of the remote creature a principle pleasure of the species engagement I attempt. How do a pet and a wild bird differ? Engaging a wild bird, birdtalking, is momentary and always terminable by the other party. Birdtalk is transient, though one may wile hours in such engagements.

Birdspace—roughly, for passerines, the ratio of thirty feet to one human yard—of course varies with species. Perhaps for an Eagle it's more like three hundred yards to one human yard. The point is, birdspace and birdtalk fit. Bird voices and bird patterns of life correspond, as do, I suppose, our American transience, mobility, and cell phones.

The snowlight gives a male Cardinal at the feeder a garish glow equal to his glissando song; I have seen wedding photographers on the Bay of Naples use large metal reflectors that give an effect similar to the snow. There are other surprises. The Titmouse has no cockade, probably concentrating too much on his or her food. I didn't know it retracted. This feisty little bird, so quick to defend its trees, here seems tame indeed. It's a bit sad, but we can intuit the meaning of "a state of nature." Winter must be got through, must be survived. For the humbled Titmouse, this snowday is not as exhilarating as any day in March.

But some birds do sing in near-freezing low-angle January sunlight. Although there are no flowers out in Kensington Gardens, London, nearby in a fenced, side-street green with access only to certain residents, Blackbirds start with high hawklike in-sucking calls. Like an Osprey, but firmer, harder. One flies from a huge plane tree behind some fourth-floor construction staging and plastic tarps. After I return the call, the bird sings a descending fifth plus a longer fourth:

Another Blackbird, I think, makes the staccato takeoff series very like an American Robin. As I continue to mimic, one starts virtuoso short trills and intervals.

Before toast (and dawn) the next morning, around six, I hear a Blackbird in the green behind my room, the one that borders the parish church of St. Stephen's South Kensington. That's the church where T. S. Eliot headed the fire brigade in World War II. Its song is unaffected by the chill:

Later, walking Victoria Road to catch bus number 10, I hear a descending minor third with emphasis or accent on the second note, *dih-dah*. Along here the trees are smaller, decorating the small but elegant yards of the two- or three-story row houses. Only twenty feet away, I see the bird making the call; it's Chickadee sized or a bit larger, in the top of a leafless tree. A black streak down its chest and belly separates the dusty yellow breast flanks. It's the same bird I heard a week ago over in Via Giuseppe Dezza, Milan. The Italians call it *cincerella,* the Brits, the Great Tit. (See art for previous chapter. Be careful about discussing this species in bars.)

Back home, in February, a high wind sweeps across the harbor bridge in my town. Invigorated, but with binoculars fogged, I have trouble identifying one of only two birds near channel rocks. Nearby a Black-backed Gull looks similar, except this one has a black chapeau on a bright-white head and neck. I suspect the wings are black, though they're half submerged. Peterson says the Black-crowned Night Heron "winters n. to s. New England (a few)" (1947, 16).

One cold morning in February I hear repeated Carolina Wren songs greet the sunrise. The lengthening daylight trumps the temperature in bird values. And, as I have mused, the Carolina Wren may well feel that under the added cold the sun requires its special matins assistance in order to rise.

The last week in February, with the temperature about to rise to fifty degrees—but could the birds know *that?*—I see what looks like a small Robin, red-breasted, in a low leafless bush. Not on the ground. I get out my binoculars and notice a blue cap: it's the first Bluebird I've seen in three years. I go inside and look up Schuyler Mathews's version of its song and also Cheney's. When I come back out, it's gone. I try the song by the book. No luck. Perhaps it is the song I heard walking home an hour ago in the village. One unfamiliar to my ear, plaintive, almost a trill. I thought it to be some Warbler. Maybe it was, but maybe it was this visitor.

Down by the salt marsh I hear Canada Geese in their usual racket of takeoff. Why do they talk so much during flight? Today, flapping and barking away they rise maybe two feet above the grass and hold that flight plan for a hundred yards until they're above water again. They keep yapping and honking another couple of hundred yards until they reach grassland; then they veer up over the near trees, still talking, and shortly fall silent. I presume they landed in a field just the other side of the junipers lining the far side of the river.

The mystery is—why all the noise? Contrast swans, notoriously mute even when not onomastically so. Geese are more gregarious, more voluble. They seem to put more effort into flying, too. Perhaps they bark encouragements, rather like the "Go-go-go-go-go!" along the ski or decathalon routes at Olympic venues. Contrast the silence of the Hawks I see perched along the highway as I drive north at dusk through a dusting of snow. Every couple of minutes I notice another one ready to feed on a hapless rabbit crossing the road. Or is he desperate enough to fend off the Crows from their roadkill? Only once at dusk was I surprised to see a Hawk feeding on roadkill, but it may have been an Owl. It was dark enough.

Living on the Atlantic Flyway and just within the holly range (read, fairly mild winters), I witness daily Canada Goose commutes. Sometimes I think our house is at the intersection of two gooseways. Truly, they are *highways,* a couple of hundred feet higher than where I stand. Gooseways never have cloverleafs. They don't even have proper ninety-degree intersections, nor stop signs. They do include "yield" instructions, by word of beak. Geese wedges, sometimes two hundred strong, always cross in *X*s. They seem to know toward exactly which point on the compass they're heading. Unwavering, undeterred by the traffic at the intersection, they flap full-on. Yapping away, beating crosswind, two geese masses head onward. From the vantage of this flightless observer, they seem to be at the same altitude. But this cannot be true. Also, their voices, their *honks,* seem similar, but this again must not be so. I still must develop my ear to tell familial distinctions that allow such masses of traffic to converge and then to proceed undeterred. In my experience only the French motorists at the Place de la Concorde operate with such straight-ahead minds. The place is a traffic circle, but the Parisians cut secants through it, like the gooseways over my yard.

You find them throughout Venice, while sitting in a campo bar, in a narrow *calle,* or crossing a *ponte.* A good Merlo—not the kind you sip from a glass but the kind you hear, the winged variety. The European Blackbird, vintage April.

Their clear call resounds through the sidewalk-sized streets.

They make the most of gardens the size of a studio apartment or bedsit. In the astonishing compression of Venice, the song of the Blackbird is a breath of freedom, a reminder that this was a republic for a millennium before the great soi-disant republican Napoleon abolished it. The Merlo's sliding glissandos often sound darkened somehow with overtones.

I am no Saint (Francis), but I do talk to birds. They respond. Merli speak in a range low enough for humans to imitate or respond. They vary a major triad, like the first three notes of the U.S. national anthem, "O-oh-say." But usually they do it backward, "say-oh-O." If you whistle that, and slur the first two, you approximate one Merlo sequence.

In mid-March at sunset on the Lido San Nicolò, I heard "say-oh-oh-oh," or *do-mi-sol-mi,* then *do-mi-do-sol.* The Merlo varied with slurred grace notes and glissandos, for example on the sol. Like most birds, the Merlo do not sing diatonic pitch. They use many quarter tones, as do several ancient world song traditions: Indian, American Native peoples, and others in Asia and Africa.

When I whistle back, I too do variations, since direct mimicry confronts and challenges. It is useful for "calling" birds but not for birdtalk with them.

The Vaporetto Linea Uno drops me at San Toma, across from the Palazzo Mocenigo (Vecchio—there are three) where Giordano Bruno was turned over to the Inquisition by his student. Talk about bad student evaluations. A short walk brings me to the great playwright Goldoni's house—actually, his physician father's, where Carlo grew up. A fine, tall courtyard stairway leads to such joys as an Eighteenth-Century marionette stage with commedia dell'arte puppets. Goldoni wrote the definitive trilogy on summer vacations, as good as Chekhov's *Seagull.*

Leaving Goldoni's place, I cross the Rio Meloni (hit the first syllable) and hear a minor triad in jazzy rhythm, first descending *sol-do,* then rising *mi* (flat) *sol. Dih DAH DAH did*, the rhythm. On a plaque I read the French ambassador to Venice built this palazzo in 1391. Largely rebuilt the year Giordano Bruno came here for the last time, in 1591. I imagine a great-great-great (this could go on) progenitor of this Merlo encamped here by the French ambassador's window. And never left.

Walking to the Rialto I pass among myriad stalls—scarfs, ties, handbags, men's shirts, women's jewelry, wooden boxes. A beautiful mahogany cart offers Signore Geppetto's wooden marionettes. He makes them in the mountains north of Vicenza. Pinocchios of any size, to fit any *borsa* (budget).

Crossing the Rialto Bridge, also built in 1591, I see to the west Pietro Bembo's Renaissance palace colored between rose and peach. It's now an insurance company. Once over the bridge, I pass the statue of Goldoni and cross into San Marco *sestiere* or district at the Cheese Fava. Its *campo* like the others is paved with stones, not a field as implied by the cognate "campus" or its Latin root. The Campo Fava was cut in centuries ago by a family with political pull—the Renaissance equivalent of the beach houses springing up all over the coast.

I hear a joyful Merlo as I pass the Zorzi palace, now the home of UNESCO. Nearby a small campo, San Severo joins at the end Calle Preti, Priests' Alley. There's a garage-sized garden and two trees across the four-meter-wide Rio San Severo. I respond to the Merlo in the tree in my best Merliano. It's a cloudy language, bubbling with dark overtones and occasional metallic rattles. Today, this one sings a series of variants on the diminished seventh chord, the diminished seventh note as high as I can whistle, A♭ two octaves above middle C.

Back in the lagoon, the *vaporetto* (boat) takes me to my final stop, the Giardini, with its busts of Verdi and other composers, also with the exhibition hall for the Venice Biennale. Again, the Blackbirds precede me. I have found a good Venetian Merlo.

When I get to London, I hear my familiar Blackbirds throughout the city, in every square or garden from Cornwall Gardens in Kensington to Lincoln's Inn Field, from Canonbury to Russell Square, from Kensington Gardens to Berkeley Square. This brings me to the famous American jazz standard "The Nightingales Sang in Berkeley Square." Most probably they were Blackbirds, which, like English Robins, do sing at night.

Multiple London birds talk. Instead of the startling, solitary Blackbirds of Venice, I hear a tumult of high-pitched, domineering little Robins, chattery Wrens, and—especially along the New River (new in 1600!)—clicky Blue Tits and sociable Blackbirds. The scattering of public squares, parkland gardens throughout the city, with their century-old plane trees combine with many formidable private hedges to make the city bird-friendly despite the decibels. A recent article claims that the birdsongs of mating have had to increase so much that were a male Nightingale to sing by your ear, it could damage your hearing. Yeah, *roight*—as if the lorries and buses and motor, the radios blaring rock and rap aren't already doing a good job of that.

Down in Dorset I wake with the Blackbirds one May morning, just before five. It's going to be a clear day, the Moon a perfect sphere over Weymouth Harbour.

I plan to go outside and notate as complete a Blackbird song as possible: I have secured several musical staffs for this purpose and my pen. But first I make tea. It's a chilly morning, and I am not awake. By the time I've drunk my tea with a touch of honey and warm milk, the nearest Blackbird is gone. When I venture into the six degree chill at a bit after 5 a.m., the bird scape has changed. I patrol the knolls at Overcombe Drive and hear very few Blackbirds. I do see Jackdaws, a Carrion Crow, Wood Pigeons, and—is it a Starling?—on chimneys and roof beams.

Birdlife is fast. Especially in a chill. On warmer mornings Blackbirds luxuriate in the sunlight and their own voice, before or after nesting ensues. Right now there's food to be searched. There are little ones, little gray bundles—one of which I saw butchered yesterday by a nonchalant Carrion Crow on a fine copper gutter. I have my ineluctable tea habit, this Crow his habit. This morning I see Jackdaws feeding on the grassy area overlooking the sea. I also see one Blackbird, with his yellow beak and yellow eyelid, at the base of a light pole. But he or she is not talking. Just collecting humble food.

After nearly an hour, back at my own place, I hear one. It's seventy yards away from my digs, on a neighboring gutter. He is talking as he did an hour ago, with his individual phrase. I engage him and respond, but quietly. He proceeds with variants, with some clicks and chirrs too. I take maybe a ten-minute lesson in bird, and I write down about a third of what he has said. As his amanuensis, I am frustrated at my own novice capacity. Yet this bird does not seem to be a genius. Even the average Blackbird is so much better at this that I am humbled.

On a misty, sunny, breezy midmorning I walk the residential knolls of Overcombe Drive. A black bird sits atop most every tall tree or power pole or roof. Some are Daws, some Starlings, and many are Blackbirds. In this neighborhood—does it extend a street, or a postal code, or a country?—they announce variants on the WWI marching song "The Caissons Go Rolling Along." "Over hill, over dale . . ." As their scales use quarter tones, sometimes the caissons roll in minor thirds, and always at the end each adds a unique fillip, either a rattle or cackle or skip.

Song Sparrow

Spring Birds

*A*h yes, spring and birds—a tautology. When *is* spring? Birds define it with their return to northern climes. Swallows to Capistrano. Robins to the Emily Dickinson homestead. So "spring birds" is redundant. With the sighting of the first Robin, spring manifests itself, although recently there have been nonmigratory Robins in southeastern New England. This year let us say spring begins with the first Bluebird, on February 20. The branches are bare but filling with dozens of bird varieties. The air rings with minor thirds in a dozen timbres: Cardinals glissing up, Titmice glissing down, Chickadees with their two sustained notes stretching almost to a minor third, rare and tentative Bluebirds, Robins still mostly silent while coursing the yard for worms, and insistent Carolina Wrens, Mockingbirds, Warblers, and high Starlings in the range of Warblers.

I feel dazed, like a symphony conductor who has lost control. I doubt that I've even identified all the various singers, much less responded to them. I do respond to a Titmouse, who then varies his call. He first makes isolated calls with perhaps a quarter note rest in between, but once I answer, he speeds up the series. When I repeat the faster version for a couple of minutes, he varies the rhythm into a rising call. The slight change sounds to me like a huge difference, from "Stay out of my tree" to "Come to my tree—so we can take it outside."

I remain about fifty feet away, under the bare branches of another tree, perhaps just outside the tree-line territory he is claiming. A couple of days later, I hear dueling Titmice, descending unslurred minor thirds, a half tone apart, as in the first two measures above. I do not enter the fray to confuse the issue.

At any rate, today is gray, nearly fifty degrees, and the myriad birdcalls fill the air with a Gold Rush, homesteading land-rush feeling. So many of these oscine passerines share similar notes in the octave above middle C; I reflect on the pond birds—Mallards, Black Ducks, and Wood Ducks. It's the Mallards who almost sound like frogs today. Nuthatches, too, have a froglike, cawing note, like a small Crow. They almost share language across species—an amphibian-passerine summit.

When I arrive at work rushed, having parked far away, I hear a clear piped descending major third. Unmistakable. A Blue Jay, out of his myriad disguises. In a month this call will lose its second note and become, for a couple of weeks, just one piped, clarion call. For years it would surprise me every spring, so un-Jay-like. Yet it is the thing itself, the "Jaynote." A couple of weeks later, in mid-March, I make this jaycall to flocking Blue Jays. They fly closer, look at me. They don't return the call. It may well be a breach of etiquette, like saying "Bon voyage" to someone who has no upcoming vacation, or "Good morning!" at dusk. One week into March, the predawn has a chaotic, festive, and battleground feeling. Titmice echo, half a tone apart, from a hundred yards away. A trifle. Normal conversational range. When is the last time you talked to someone a hundred yards away? I'll bet you were on a boat. Or in a parking lot. It's remarkable, really. These little creatures, the weight of my little finger, assert themselves across—indeed, *occupy*—acres. With them are a Cardinal not yet warmed up and a Song Sparrow calling from far away, down by the open fields.

March 9 this year I talked to two Song Sparrows on my college campus. They were in two pathetic leafless trees planted last year; but they were near the water, and the song (only one sang) rebounded from the brick walls. They did not discuss their migration, although I asked. (I asked by trying variants on their song, hoping one would be southern dialect.) What had they seen? Long Island? The Carolinas? They spoke of now, of the day, of their expectations. As if drunk with the sun, they spoke of their fond hopes, as if wars, crimes, and human evil did not exist.

A day later, another Song Sparrow dueled from thirty yards away. I couldn't see him, but I know the song. The initial rhythm is the same as the prior—"prior" only because of its location, closer to where one was sighted earlier. I couldn't swear they are the same song or the same banded bird. A rough approximation of the duel goes like this:

The second ends with an in-sucking sound, here shown with discord. Both are quite different from the Song Sparrow between fields near my house fifteen miles away. No two Song Sparrows sing at all alike except for the initial rhythm.

Meanwhile, the Cardinals are back. At first I confuse their initial shortened whiplike glissandos, rising minor thirds, with those of the Titmice, usually a bit lower. The Titmice's rising, slurred minor third is their invitational rather than their confrontational (descending, more staccato) call note.

They sound similar to an untrained ear, but I believe they have entirely different meanings. I have never heard the rising, invitational one used in a close-up battle for territory, whereas I have heard the descending, unslurred one used many, many times, often by two Titmice on contiguous branches.

The Cardinals, like the Jays, use their spectacular coloring to attract attention; the Cardinal song can attract plenty of attention, too. For the same effect, the Mockingbirds must jump and hover over topmost branches to display their white wing stripes. In early spring, the Cardinals leap in song intervals, fourths and fifths, even octaves. Here I heard one slow down his song later (after the double bar).

On the Ides of March I see one perched halfway up a tree, not on top, making a descending triad, repeating it a couple of times, and then adding the usual whiplike glissandos. Shortly afterward, he flies down low (altitude three feet) toward my house and into shrubbery. Perhaps a nesting site. They have nested nearby in previous years. *Pyracantha,* you know, provides plenty of thorny protection.

Such a variety of similar songs among the Cardinals; as I leave my ATM one day I hear an unusual one, perhaps immature, or perhaps a Cardinal idiolect. Instead of the usual triplet, there's a catch or a pause or a thirty-second-note rest between the first two slurred notes and the third (a harmonic fifth):

I go back into my small bank and tell the teller; she says she'll listen for it when she leaves in seven hours. With such ease we have all adapted to the human-made environment. Yet, look at what we lose—what *she* loses—in our disconnection from the natural or the wild.

Gray Day

One rainy morning a week later I talk to a Robin outside the small, shingled takeout bakery in town. Misty, the rain has paused right now. Lots of English Sparrows *cheep, cheep* in the nearest scraggy tree. I hear a few Robin plosives, which I wait to imitate because they are bird warnings—the kind of call they make when they spy a cat. Then, as in the prologue to this book, I purse my lips, but instead of in-sucking a *cheep* (also an avian warning), I burst one staccato, plosive note, tongued as if I played a bugle, flute, or trumpet. Ten seconds later I make another. Not sure of the effect of my returned warning, I see the Robin perhaps thirty feet away, making a couple of more plosives, then flying in back and above the little bakery. Now maybe fifty feet away, above the bakery, it starts a song. But instead of the usual burbling plenitude of alternating triplets and higher glissandos, descending, I hear a minimalist song—two glissandos, a fifth leap between:

I am amazed, almost astounded. Never have I heard such a reduced Robin song. Is it the gray, rainy weather? Feeding Robins often sing more strongly in rain. Could it be a female? In the poor lighting I don't really notice. What I do observe is warnings, lots of them. Over the course of the whole rainy day (later listening to the birds a half hour away in Providence) is the full panoply of "Watch it!," "Careful, there!," and *"Attenzione!"* All in Bird, like so: *chk chk, cheep, chp, tssk, sskp,* and then the Robin's warning note I made earlier. Robins repeat that sforzando plosive in their flight song on takeoff, each note rising a half step (or less, a quarter tone) in a series—often of five. (See the rising chromatic call of the London Blackbirds, above in the last chapter, "Winter Birds.") Today, I hear this call quite often, and all the assorted warnings of many bird species.

Are birds more threatened, more vulnerable to predators, in dark, gray weather? It appears so, on our evidence of one day late in March. Probably their enemies—feline, avian, and herpetological—all count on more success with reduced visibility. There are, of course, proverbial shades of gray. Today, the day after, is quite bright, though still overcast. I hear fewer warning notes, fewer raucous Robin flight songs. Perhaps, too, they can't use their own bright male colorings so easily to distract when all colors are reduced. And what do we know of bird vision in the gray range?

A week after that minimalist Robin, on another gray day, I hear two other birds using the same intervals. It may be coincidental, or it may be that short glissandos a fifth apart are the right calls for this leafless, gray April the second. At 5 a.m., predawn with only the dark outlines of buds and branches, I hear a Cardinal twenty-five feet from my roof-window. It's cold, maybe thirty-five degrees. Because he's close, I subdue my response, just the first two notes. To my weak notes (see after the double bar), he truncates his song as if listening. Probably he is. He reduces his song to triplets.

Ten minutes later, a Robin makes a burbling version of these same notes (the first two measures), with a few more sliding brief glissandos. Finally, I hear the same notes, now an octave lower, slower, and with such a different timbre or tone quality.

This is the Mourning Dove, like Cardinal or Titmouse recorded at 78 rpm and played at 33. I don't know what to make of these shared vocal gestures, shared tonalities, even. But surely they hear each other's similar phrases and know they use common notes. Of course, each bird's note—say F, an octave above middle C—made by Robin or Titmouse or Cardinal or Mourning Dove is distinct and recognizable, usually.

Spring birds, like flocking birds, use silence, as well. The silence of nesting birds, and especially nest-building birds, is profound. This same dark, gray April 2 I opened my storm door to return a Robin's short "reports." It's dark, near freezing. I don't have my coat on. After fifteen seconds with no response, just before I close the door, something dark rushes from near the porch, almost under my feet. It's a Robin, perhaps the mate of or even the very bird I was talking to. It had fallen silent, searching perhaps for a nesting site. The song may have been simply to attract—and to distract—other creatures, as it had me.

A couple of minutes later I saw first one silent Robin, then another, dark apparitions in a nearby apple tree. Neither was singing or had sung for a while. Their business, their mission, required "radio silence." Often in nesting season a male Robin or Cardinal will sit on a top branch and give a spectacular display:

Meanwhile, the female flies low, entering or leaving the nest or nest-to-be. Having completed his distracting display, the male too flies low as the female did previously.

In the midst of their song, birds are their birdiest. Completely full of themselves and their music, thinking only of—rivals of the same species? They direct their song at other birds. They enact "speech acts," gestures that hold meaning for others of their species and possibly other bird species. At their birdiest, songbirds remind us of ourselves. Most all human endeavor centers on other humans. And how limiting such a perspective is. Would Napoleon have invaded Russia because of something the sturgeons did? Or the yaks? Has any human ever declared war on the pets of another society? Although the consequences of war inevitably devastated the enemy's domestic animals, usually they were not targeted. Let us be thankful that our species-blindness has at least this one minor blessing.

Early April in a cold snap, the wind is up. Where do they go, the birds? Closer to the ground, doubtless. In the low forties today; last night I observed a dozen gangly Great Egrets, big as Blue Herons, heading toward the pond. They pulled up just shy of it, in the tops of trees. Quite a sight, like Jason's golden fleece, but these were large white "fleeces" perched in medium-sized trees a hundred yards across a small pond. At five o'clock this morning they were still there, but gone by sunrise. Walking by a larger pond this morning after very brief snowflakes, temperature in the low forties Fahrenheit, I see a snaky, dark neck and up-slanting, long bill. Cormorant, its body entirely submarine, as it often is. Then it "rows" (as Dickinson put it) in the air, paddling with its feet to get airborne as I do in my dreams. Fifty yards farther out a small duck-like critter pops to the surface. Using my binoculars I determine it's a Puffin just before it somersaults back into its feeding medium. None of these birds are talkative, though the passerines still repeat their calls in the cold. An exuberant Cardinal celebrates the dawn, along with a chorus of Titmice, Chickadees, and Song Sparrows.

I'm amazed next day in the wintery forties, by how much energy the Titmouse shows with his constant rising major-third glisses and occasional more rapid descending minor thirds, often a half step higher. He "covers" my yard and a neighbor's eight-acre field. When he's on the far side of it, I start my rising-third response. Although I can't see him over the knoll, with every response he sounds closer. And then he appears. I vary my response so he won't think I'm challenging him. I decide to try a little Mendelssohn on him, the Italian Symphony. *Da-deed-da da-deed-da* (his major-third call) to which I add the fifth.

He responds with his same two rising glissandos. I do it again. His same again. But on the fifth or sixth try, after his two glisses and my Mendelssohn sound bite, I hear just the rising gliss from the third to the fifth. He did it! A joyful moment. He also flew back farther in the field. A mini-good-bye. Though he did not comprehend the whole phrase of Mendelssohn, he did understand an additional tone. We shared a musical idea.

Simple mimicry is not conversation, though it may act that way among birds parodying each other. Conversation requires saying something other than what is said to oneself. To converse with a bird, one must vary what the bird says. Recently, this spring, I began varying my imitations by a half tone, or with shorter glissandos. I think that the less public, more domestic birdtalk—say, between mates—employs shorter glisses, as well as quieter.

I end this section on spring with my homemade poem on spring's transition to summer. I give it a French title after a character in Shakespeare's *The Tempest*.

Lai d'Iris

Let the mouse-ear hawkweed dare
To challenge daisies, and appear
When the leaves their yellow trim
Sprinkle on the trunk and stem
Of wild cherry, flowering pear
And crabby apple. Let the year
Begin in strife, pugnacities
Of Wren and Warbler burst in spurts
Of bird barks. Let Robins fill
The gorse with nests, and Cardinal
Choose white-blossomed pyrocanthus,
Whose orange berries in the fall
Splash surprises where they will.
With the summer's midday yawn
Let the lilacs turn to lawn
And iris, second-yeared, grand-dam
Of the fleur-de-lis. Then butterfly-
Weed open like a symphony,
Bright with bees and those that name it,
Moth-sized, buying golden pollen
With their labor, dusted their wings
With it—or is it just the way
Themselves are dappled? And by now
The ox-eyed daisies follow purple

Loosestrife, joe-pye weed and dock,
Berries manifold, black or wine-
Berries, raspberries gone by,
But blue (wild ones) on mountainsides
In Richmond next to Tanglewood.
Now August's spiky thistle, sting-
Ing horse nettles, and rods on rods
Of goldenrod—the steeple kind, or
Cantilevered, the Thai dancer type—
Nodding, waving. These I've seen
Without a name; now they come
When I call them to this poem.

Cardinal

Animal Language
The Linguistic Debate

*D*o birds, or any animals, have an actual language? I have proceeded on the tacit assumption that for our purposes they do. I do this first to contradict the usual human-centered view. Second, it is a central assumption for interacting with birds as we have. But at this point it should prove useful to look at the historical discussion of language in both mankind and animals. In Noam Chomsky's *Cartesian Linguistics* we learn that Seventeenth-Century French thinkers such as Descartes, Cordemoy, La Mettrie, and Bougeant set the terms of this debate. In the next century the grounds of the debate shifted to German theories of the origin of language by Romantics like Herder and Schlegel.

Père G. H. Bougeant's *Amusement philosophique sur la langage des bêtes* (1739) suggests the tenor or register of the discussion. Bougeant states several ideas that recur until this day, for example, *"tout le langage des Bêtes"* (all the language of the beasts) reduces to expression of their sentiments and passions. Nor are those passions numerous. Following Locke, Bougeant argues that animals have no "abstract and metaphysical ideas" In a passage that seems to describe Hartshorne's "non-versatile" songbirds, his monotonously repetitive class of birds, Bougeant says, "*Il est nécessaire qu'elles répètent toujours la même expression, et que cette répétition dure aussi longtemps que l'objet les occupé.*" (It is necessary that they keep repeating the same expressions, and that such repetition lasts as long as the object occupies their mind) (Chomsky 1966, 11).

Not only do they repeat the same expression in the presence of a specific stimulus, but they cease to repeat it when the object no longer strikes their senses: "*Elles n'ont que des connoissances directes absolument bornées a l'objet présent et matériel qui frappe leurs sens*" (Chomsky 1966, 11). Many modern ornithologists, such as Alexander Skutch, also attribute birdsong to immediate sense data and their resulting affects.

Birdsong at dawn, under Bougeant's theory, results from the sighting of auroral glimmer. Skutch agrees but adds that song results from heightened emotions. I have speculated that there may well be an abstract or "metaphysical" addition to these, that birds act as if their song helps the sun to rise. The more they join in and participate, the more albal success they have. Most do not sing anywhere near as much—as often or as long—on overcast days. On a windy, cold March day there is virtually no birdsong, while yesterday, in the sixties, they were legion. Toward dawn, it's as if they begin cheering, praising the sun, and as it gratifies their praise and cheer, they sing more.

John Locke raises the issue of bird language a half century before Bougeant in *An Essay Concerning Human Understanding*. Under his discussion of simple ideas, and particularly their retention, Locke says:

> This faculty of laying up and retaining the *ideas* that are brought into the mind, several *other animals* seem to have to a great degree, as well as man. For, to pass by other

instances, birds learning of tunes, and the endeavors one may observe in them to hit the notes right, put it past doubt with me that they have perception and retain *ideas* in their memories and use them for patterns. For it seems to me impossible that they should endeavor to conform their voices to notes (as it is plain they do) of which they had no *ideas*. For though I should grant sound may mechanically cause certain motion of the animal spirits in the brains of those birds whilst the tune is actually playing; and that motion may be continued on to the muscles of the wings and so the bird mechanically be driven away by certain noises, because this may tend to the bird's preservation: yet that can never be supposed a reason why it should cause mechanically, either whilst the tune was playing, much less after it has ceased, such a motion in the organs of the bird's voice as should conform it to the notes of a foreign sound, which imitation can be of no use to the bird's preservation. (Locke 1.122)

Locke appears to refer to some account of bird learning, perhaps the same by the "most learned Dutchman" referred to in Thomas Fuller's preface to *Ornitho-logie* or *Speech of Birds*. Fuller says that the Dutchman "hath maintained that Birds doe speake and converse one with another: nor doth it follow, that they cannot speake, because we cannot heare, or that they want language, because we want understanding"(1655, A2).

More likely Locke refers to some native English experiments with birdsong, for he concludes specifically in regard to bird imitation of nonbird sounds, sounds that were in fact produced on the previous day. This seems very like the kind of naturalist intervention reported at the new Royal Society.

> But which is more, it cannot with any appearance of
> reason be supposed (much less proved) that birds, without
> sense and memory, can approach their notes nearer and
> nearer by degrees to a tune played yesterday; which, if
> they have no *idea* of in their memory, is now nowhere,
> nor can be a pattern for them to imitate, or which any
> repeated essays can bring them closer to: since there is no
> reason why the sound of a pipe should leave traces in their
> brains which not at first but by their after endeavours
> should produce the like sounds; and why the sounds
> they make themselves should not make traces which they
> should follow, as well as those of a pipe, is impossible to
> conceive. (Locke 1.122)

I can't claim in all my interactions with birds that one
ever reiterated a tune produced yesterday better the next
day. The closest I came was in "teaching" Beethoven to an
Oriole. He rejected the final cadence, nor did he improve the
next day. Having said this, I must admit the avian training
Locke intimates suggests Tchernichovski's experiments with
mechanically programmed plastic birds (see "From Guns to
Birdfeeders").

Locke's most famous passage on birds concerns a parrot
in Brazil. Like most discussions of animal language, this one
focuses on the bird's learning human speech, rather than our
learning his—as we endeavor to do. Soon after his discussion
of metempsychosis, Locke regales his reader with the story of
Prince Maurice.

> I had a mind to know from his own mouth the account
> of a common but much credited story, that I had heard so

often from many others, of an old *parrot* he had in *Brazil* during his government there, that spoke and asked and answered common questions like a reasonable creature, so that those of his train there generally concluded it to be witchery or possession; and one of his chaplains, who lived long afterward in *Holland,* would never from that time endure a parrot but said they all had the devil in them. (Locke 1.279)

Locke assesses what one might call the courtroom demeanor of the witness, "[Prince Maurice] said, with his usual plainness and dryness in talk. . . ," that there was some truth to it, that he had asked upon his arrival in Brazil the parrot be sent for, though a great way off. The bird turned out to be very large and old:

> When it came first into the room where the Prince was, with a great many *Dutchmen* about him, it said presently: *What a company of white men are here?* They asked it what he thought that man was, pointing at the Prince. It answered, *some general or other.* When they brought it close to him, he asked it: *D'ou venez-vous?* It answered De Marinnan. The Prince: *À qui êtes-vous?* The parrot: *À un portugais.* Prince: *Que fais-tu là?* Parrot: *Je garde les poules.* The Prince laughed and said: *Vous gardez les poules?* The parrot answered, *Oui, moi, & je sais bien faire,* and made the chuck four or five times that people use to make to chickens when they call them.

Quite a little conversation, about what strangers always ask: What do you do? I watch the chickens, *cluck, cluck.*

Now, this whole conversation was originally in "Brazilian"—not clear whether Locke means Portuguese, or dialect, or a native language. One would assume that since the parrot belongs to a Portuguese man, he would have taught it Portuguese. Locke continues:

I set down the words of this worthy dialog in *French,* just as Prince Maurice said them to me. I asked him in what language the *parrot* spoke and he said in *Brazilian.* I asked whether he understood *Brazilian;* he said no, but he had taken care to have two interpreters by him, the one a *Dutchman* that spoke *Brazilian,* the other a *Brazilian* that spoke *Dutch;* that he asked them separately and privately, and both of them agreed in telling him just the same thing that the *parrot* said. (Locke 1.280)

Locke concludes that the "Prince at least believed himself in all he told me, having ever passed for a very honest and pious man." He leaves it to naturalists to reason, and to others to believe as they please. For our purposes, the parrot's ability to mimic human speech, even on a "baby Greek" level, is less to the purpose than its owner's ability to communicate with the bird. Did it ever confide where its enemies roosted?

The mid-Seventeenth Century saw the growth of inquiry in natural philosophy—what we now call *science*—such as aberrations at birth, or "monstrosity." With this study of what we now call *genetic defects* came discussions of what defines "man." Locke cites the example of a French abbot, Mon. Ménage, whose defects at birth stymied the priests. They wondered if he should even be baptized:

> Nature had moulded him so untowardly that he was
> called all his life Abbot Malatrou [i.e., ill-shaped]. . . .
> This child was very near being excluded out of the species
> of man, barely by his shape . . . but no reason why a
> rational soul could not have been lodged in him; why a
> visage somewhat longer, or a nose flatter . . . could not
> have consisted with such a soul. (Locke 2.56)

What bodies may a soul inhabit? This impinges on our subject if language is a principal evidence of mind or soul. (For the Greeks they were both *psyche*.) The idea of animal languages then, like pet cemeteries, treads the edge of heresy, as it also stretches the limits of compassion. For instance, it turns out that the complex social life of ants includes a language, partly stridulation, a certain squeaking. And ants can hear caterpillars, in their own interspecies encounters (Milius 2000, 93). But the language of insects defies human sympathy. Nor do the ancient Pythagoreans or Hindus, as far as I know, include ants in their hierarchy of souls.

Birds, on the other hand, keep appearing as symbols of the soul in the Western tradition and apparently in some non-Western traditions—Native American and Asian. An example from English literature arises late in Andrew Marvell's famous poem "The Garden" (see "Literary Birds," p. 92). Paradoxically, to claim that birds have souls is heresy in orthodox Christianity. At the same time, birds are common symbols of the soul in that same tradition.

Chickadees

Birds of America

$\mathcal{R}$ock music fans tooling along in open-windowed Camaros, motorcyclists on BMWs and Harleys, women mowing lawns, motorboaters, suburban men on riding lawnmowers, moped commuters, golf-course greenskeepers with weedwackers, joggers with earphones plugged into cassettes or CD players, joggers even without earphones (too exhausted to listen), residents near airport landing strips, people who live along the flight paths of smaller corporate and private planes, riparian residents when the hydroplane is running, the same bankside residents when motorboats grace the river, coastal dwellers during high surf, everybody on a stormy night, most everybody on a stormy day, cyclists with earphones and CD players, longhaul truckers in big rigs, cyclists whose own thumping hearts drown out everything, local delivery van drivers, all motorists with car windows closed, most motorists with them open, trash collection drivers and workers, those with ear damage from rock concerts, those with ear damage for any other reason, and . . . the deaf. These are some of the people who can't hear the birds. They can't hear them, or they don't listen, or they can't be bothered. For many of them, birds are trivial, their songs negligible compared to a rock song.

Even if they wanted to hear the birds and, say, the peepers, the frogs, the crickets, the squirrels, and so on, it's getting harder to do. In an interview broadcast on national news, Bernie Krause, who records natural sounds professionally, estimated that when he started out he needed ten hours to come up with a one-hour recording without motors, jets, and other human sounds. Now it's more like two hundred hours to one. His interest is *biophony,* the natural panoply of sounds across the wavelengths that enable the various natural necessities—frogs camouflaging themselves in cacophony that distracts predators, scarce birds finding mates (Krause 1998, 3; Beebe 1983, 5). Our mechanical and human sounds seriously impair this basic order, or disorder.

For those who could not care less about birdsong, they themselves—and since the list once included me, *we*—have much in common with birds. We Americans are constantly on the wing, moving long distances in a single day. To work—an average commute of twenty minutes to half an hour—and home again. To the store. Large proportions of our Northeast residents migrate south for the winter, to Florida or the Southwest. They have even acquired the appropriate name of snowbirds.

When we see flocks of starlings moving en masse from power line to rooftop and back, or two flocks of Canada Geese crossing in some imagined interstate interchange in the sky, or even Sandpipers scurrying on the beach at the water's verge, don't we see our own behavior, too? What are our daily commutes but short flights. And for the last fifty years we have been flying all over—literally. Air traffic is nearing gridlock. Where is everyone going? Where we are not. What are we looking for? Food and nesting sites—and a lot that birds don't need. Better plumage. Better interest rates for retirement accounts. Airline tickets.

Still, the fundamental comparison of bird behavior and

human behavior, especially with regard to movement, is probably more true now than at any time in human history. And most true in the United States. Most of us are involved in massive movements on a daily, seasonal, and yearly basis. The hectic life of the suburban SUV mom, picking up this child from school and dropping them at a lesson (say, tennis), or organization (say, Boy Scouts), driving on to child number two's school—and so on. What is this but fledgling tending, following the brood through the undergrowth, often done by male bird parents, while the female nests the second batch. For most of the birds I mention, fledgling tending lasts no more than a couple of weeks, though at that stage the young will continue to flock with their parents for migration. And in certain cases, year-old young will return next season—that is, after the long haul south and then north again—with its parents. A few springs ago I saw a Bluebird group of three, apparently mates and offspring, check out my Bluebird house. They did not stay. It was only a one-bedroom, studio birdhouse.

I wonder if that third Bluebird, which I take to be last year's offspring, is male. In Italy this would be the *mammone,* the mama's boy who refuses to leave the nest, even into middle age. In the States, the male offspring still in home residence through his twenties is more likely to be a rock musician. Several of my successful friends with large houses have such resident offspring, usually sons. They are the singer-sons, as I noted with songbird gender. They are the Bluebird young, the songbirds of our culture.

Of course, other flocking birds, such as Canada Geese, also migrate with their offspring. Comparisons can be made here, too. As Konrad Lorenz showed, most Geese species are monogamous. The Canada Goose, throughout the menaces and confusions over long distances of migration must well benefit from the close affiliation of such pairing. It may well benefit the flock as a whole; with such large flocking birds and a near-even balance in sex ratio, might not annual or seasonal turmoil over mates really disrupt a species' ability to survive?

Christmas letters from old friends make me think of nothing if not avian behavior. "Daughter Sammy's in the Northwest, but she'll be flying to join our son in Germany. We'll all meet up at Dusseldorf." How did we as a nation get so compulsively airborne? Everyone has a different story leading to the same cumulative result. For some, army service took the daughter or son overseas to Italy, to Germany. Others studied languages, which they improved by one of the many international study programs now available. This in turn led to more extended stays. For some, exchange students initiated the fatal attraction of distant cultures. For others, it was family—and not just in one country. Often the father's family is from Italy, the mother's from Ireland. Finding one's roots can be a lifetime occupation. It involves lots of travel, of flying around. Whatever one's own roots, the root of *immigration* is *migration*. The birds know about this.

We sapiens certainly dreamed of flight long before we could do it. Leonardo da Vinci's winged man is a recent example, "recent" in terms of the history of flight. Swinging from trees is not too different from flight, gravitationally. It's a beautiful, sinusoidal series of arcs. Probably the oldest part of our brain controls flight in the sense of fleeing. What is your own personal border between flight distance and attack distance? When is someone in your space? Of course, it depends on whether we're at work or in our car or at home, on the sidewalk in a crowded city, on wide sidewalks in Denver or narrow Roman ones in Pompei. Stopping at an automatic toll booth to fling some quarters, I saw an attendant three feet away waiting to cross behind my car. I said hi. He didn't respond. Why would he? If he were in the booth he would have. But caught "in flight," he was not official greeter for the turnpike. He had no "flight song." Some birds—many birds—don't.

We're talking about Walkmans and car radios and walkie-talkies or cell phones with speakers always on—all the paraphernalia of human audio. Surely cell phones, like airplanes, bring us closer to bird behavior. Now, like birds, we can emit on a regular basis territorial calls. I'm here. Who's out there? Where are you? This kind of interchange was never necessary prior to mobile phones. You knew when you called a home phone or an office phone exactly where your interlocutor was. When calling someone at their house, you'd be crazy to ask, "You at home?" Or when you called someone at their office, "You at the office?" Not likely. If you've been to their house, you can picture the exact room, even what's on the wall, when you call someone's home phone. Same goes for an office phone, if you've been there. Of course, you rarely do picture such locations (unless you're in love). Location was not the principal content or signification for most phone calls prior to cell phones. Now, like the birds, we talk of *where*.

This morning, for instance. Gray and warm, early August. I hear a glissando, rising. It's a Cardinal, though this call is different from the Cardinal songs I heard yesterday. Whiplike, it glissandos up a fourth and then *poom!* It accents the beginning note again. I repeat the same call. It repeats the whiplike glissando but not the *poom*. What's going on? Possibilities: The glissando says, "I'm here, in my tree." The *poom* adds, "Alone." When I join in, it drops the "alone." That's possible but not probable. Maybe my *poom* was unconvincing or otherwise incorrect. Perhaps the *poom* signifies "at some distance," and when I answered, the distance was less, so the *poom* wrong. Perhaps the *poom* functions like a proper name in sapiens: "I am here in my tree, I the Cardinal Giuseppe." This is least likely of all; I am mixing several types of human signification, including individual identity, rare even in human cultures. Are any of these right? I don't know, but I do know the songs are very different on a gray morning in August from the songs on a sunny morning in May or June.

Why? Begin with chronobiology, the time element in biologic functions. Mating, for example, has strong chronobiological parameters. Most species mate in spring. Territorial selection usually relates to mating as well as feeding. Different songs exist for each. Then consider the anxieties of birdlife: Where are the rest of its species? Will the sun rise? Will a Kestrel or other avian predator show up, swift and silent? Will a snake or Jay dine on its embryonic young in their shells? Birdsong may express these anxieties, though it doesn't seem to, except in their *keck-keck* warnings.

While most writers divide songs or vocalizations into spring and fall (or winter), in fact, these are better divided into "lengthening-day" and "shortening-day" songs. Recall the English Robin, whose "spring song" begins in late December, just after the shortest day at the winter solstice (December 22). Consider also all the new songs from the "same old" birds, the ones we hear in late July or early August, *chick-a-dee* being one. Previously, the Chickadee is a clear minor third. The distinctive minor thirds of Chickadee and Titmouse and Oriole differ in timbre or tone quality, like the difference between an oboe and an English horn, or a French horn and a fluegelhorn playing the same note. Or Ray Charles and Paul McCartney singing "Yesterday." Anyone can tell the difference, but you have to have heard them before.

Compared to many elaborate in situ recordings and scientific publications on bird vocalizations, my method—if it can be called that—must seem amateur indeed. Since the root of *amateur* is "lover," from the Latin *amo, amare* (to love), I readily accept that I am a lover of birdsong, an amateur. But I would argue that my approach to understanding birds, my attempt even to respond and communicate with them in some rudimentary way, may serve as a model for the grander destiny of our species. About the time Shakespeare began writing (and parenting his first child), Giordano Bruno claimed that all the stars in the night sky are really like our sun, that moreover they probably, like our sun, have planets around them, many of which are inhabited. For this idea, which he called the infinite number of habitable worlds, and for seven other, more common heretical ideas, he was burned alive in Rome's Campo de' Fiori. The day was February 17, 1600.

Fast-forward to our Twenty-First Century. Pick up the *New York Times,* scan the front page. Several times in the past few years a news piece has appeared on the discovery of stars whose motions suggested the presence of planets, some of them huge, many times the size of Jupiter. Giordano Bruno's heretical perception is still front-page news. No inductive evidence of intelligent life on other planets yet exists; nor did Bruno proceed, as his successor Galileo did. But the existence of the millions or billions of galaxies, each with millions of stars, surely attests to Bruno's insight. And when we meet up with these creatures from other worlds, how will we proceed? We can use our earthly encounter with other species as a model. Our first communications are bound to be on the order of call notes or dial tones. They will be locative, as with cell phones. "We are here. Where are you?" They will also be call notes in that they will claim Earth as ours and no one else's. Stay out of our tree.

Birds, Souls,
and Aliens

*T*he cosmos fills with sounds, each planet in our solar system bellows, howls, and hisses, and beyond the heliopause, the plasma waves detected by *Voyager 1* hum. Speaking with birds may lead to greater things, like speaking to alien species on other planets. Engaging the world of birds may teach us how to encounter yet other worlds. No, don't beam me up, Scotty; this planet's fine. And I don't plan to communicate beyond our humble planet, though I listen to the wind on Mars recorded by the rover *Perseverance*.

Still, we aspire to find other inhabitable worlds; we look for remnants of species, even one-celled ones on Mars. In 2013 I spoke at Harvard-Smithsonian Center for Astrophysics, where I cited the University of Auckland's estimate of one hundred billion planets in the habitable zones of their stars in the Milky Way alone (Powers 2013).

My sixth-grade class coddled us high achievers, and we ground a six-inch reflecting telescope mirror, in the basement of a classmate's house, Chucky Downton's. His engineer uncle assembled the tube with a couple eyepieces. Loved looking at the moon, but never enjoyed astronomy because in New England you had to stand outside on the coldest nights for the best viewing. So I married a pro, who in college joined the same four-college astronomy classes as the man who fixed the Hubble, Jeff Hoffman. I began listening to Wood Thrush camping in the Amherst College woods below where JFK's helicopter landed when coming to honor Robert Frost: "A nation reveals itself also by the men it honors, the men it remembers." In Native American lore, Wrens and Wood Thrush outfly Eagles under their wings, David to their Goliath (Taylor).

Birds surmounting the terrene have represented deities and accompanied souls to the Other World, if not other worlds. In folklore, birds "flying between the earth and sky are associates or messengers of deities." Doves are associated with Astarte, Aphrodite-Venus, and the Holy Spirit; Ravens are Odin's familiars in Norse mythology; Cranes are both the sacred bird of Hermes-Mercury and the Celtic bird of the Moon; and Eagles are familiars of Zeus-Jupiter. These four are among the many species assigned such roles in myth and legend (Warren-Chadd and Taylor).

"Certain birds, especially doves, represent the souls of the departed. Historically, there has been a widespread belief, call it superstition perhaps, that birds are both 'psychopomps' carrying souls to the next world, and are also representations of the dead. It persists today." Carrier Pigeons, indeed, carrying souls.

When not seen as a soul, but a deity, Doves have stood for the pagan Venus and for the Christian Holy Spirit. Blackbirds—especially Crows and Rooks—are omens of death in various cultures. They do clean up dead animals, as do the wonderful flyers, Turkey Vultures. How did black birds come to occupy the powerful place of evil? A universal mythology claims "every black bird was originally white, but was stained or singed in some moralistic mishap" (Taylor 1988). "Black birds—especially corvids—are evil, think Ravens (sniffing out the dead, plucking out their eyes), or Crows and Rooks (both omens of death in various cultures). The Magpie, being considered half black (though in reality an iridescent blue black and white) was traditionally cursed for failing to go into full mourning at the Crucifixion but could be good or bad in English folklore, depending on numbers (one for sorrow, two for mirth, etc.). "It was a bird of the underworld in Germany and ridden by witches in Scandinavia" (Warren-Chadd and Taylor 2016).

According to the Maori, every brown bird was originally brightly colored but lost brilliance as a punishment or sometimes for self-sacrifice: "The kiwi is an example of the latter—according to Maori legend it gave up flight and colorful plumage when it agreed to live on the dark forest floor and eat up all the insects that were killing off the trees. For its generosity, it became the most revered of all birds" (Warren-Chadd and Taylor).

"The David and Goliath appeal of a small bird out-flying an eagle by hiding under its wing appears in several cultures. This is commonly the wren but also, in Native American mythology, the Wood Thrush (*Hylocichla mustelina*). It may be inspired by natural behavior." The Wood Thrush also boasts the best North American birdsong, pentatonic, like the black keys of a piano. For Australia I might concede the Pied Butcherbird's modernist notes (see Taylor 2017).

Rachel Warren-Chadd laments fear of birds: "A bird's cry at night can also be haunting. In North America the relentless chanting of whip-poor-wills was thought to foretell death or disaster.* In Britain Tawny Owls (*Strix aluco*) may be heard calling more persistently at night around Halloween, which a

---

*In 1968 Paul McCartney changed this for my generation with his song "Blackbird."

lively imagination could perceive as ominous (owls are widely associated with haunting) but is in fact, because they are re-establishing their territories as the young birds disperse from their breeding grounds."

Marianne Taylor notes that "owls can already fly silently through a pitch-dark forest without hitting any obstacles, can detect a tiny mouse moving about under a foot of snow using just their hearing [this book's emphasis] and turn their heads to look directly behind themselves—all abilities astounding to us. So why not add a full suite of psychic powers to their CVs? Sadly, such superstitions remain so strong today in some countries that the birds are routinely persecuted and killed" (Taylor, 2018).

Owls have better vision, stronger vocalization, and better hearing than we do. They are the most effective tool-less predator. Where would we be with just our teeth and fingernails? Perhaps God intended Owls to rule, and they have ruled the night, until night-vision scopes—again, our need for a tool.

Roosters top English church steeples. Cockerels featured on church spires in UK. Online, "Pope Nicholas I, ordered that every church in Europe should have a rooster on its steeple as a reminder of Jesus' prophecy that the cock would not crow the morning until Peter denounced Him three times. Since in most churches the steeples had already a wind vane, the highest point to place the rooster was at the top of the weather vane" (Warren-Chad and Taylor).

We end with my youthful fascination with the *Relations des Jésuites del la Nouvelle-France*, a chronicle of Jesuit missions in New France, from which I translated sections into verse. I first read *The Jesuits in North America in the Seventeenth Century* by Francis Parkman and then went to his sources. Marianne

Taylor notes, the Jesuits are crows. In her book *The Reason for Crows: A Story of Kateri Tekakwitha,* Diane Glancy tells the story of a Mohawk woman, who had been nearly blinded by the smallpox epidemic that killed her parents. A Christian, she had to move from her village along the Mohawk River and died by age twenty-four. Glancy combines her displacement with that of the Jesuits, the crows.

I versify Pere Brebeuf, St. Ignace, Canada (March 16, 1649):

> *Transivimus per ignem et aquam*
> *At eduxisti nos in refrigerium*
> *Through icefire and water, Lord . . .*

> *"To be about to have been eaten"*
> *so expressible in Latin,*
> *Moritoori lore loo-rye*
> *dee-dee-dye, die, die,*
> *We're about to have been eaten,*
> *masticandum masticandi*
> *to be crunched. What's for lunch?*

Addendum

In "The Titmouse" Emerson praises a Chickadee whose bravery saved his life when he was walking miles from home in winter (Emerson 1885, 195–98). At the end, he hears the Chickadee in spring teaching Julius Caesar's most famous words in Latin: *veni, vidi, vici*. I came, I saw, I conquered. But the poem begins in winter:

The Titmouse

You shall not be overbold
When you deal with arctic cold,
As late I found my lukewarm blood
Chilled wading in the snow-choked wood.
How should I fight? My foeman fine
Has a million arms to one of mine:
East, west, for aid I look in vain,
East, west, north, south, are his domain.
Miles off, three dangerous miles, is home:
Must borrow his winds who there would come.
Up and away for life! be fleet!—
The frost-king ties my fumbling feet,
Sings in my ears, my hands are stones,
Curdles the blood to marble bones,

Tugs at the heart-strings, numbs the sense,
And hems in life with narrowing fence.
Well, in this broad bed lie and sleep,—
The punctual stars will vigil keep,—
Embalmed by purifying cold;
The wind shall sing their dead-march old,
The snow is no ignoble shroud,
The moon thy mourner, and the cloud.

Softly,—but this way fate was pointing,
'Twas coming fast to such anointing,
When piped a tiny voice hard by,
Gay and polite, a cheerful cry,
Chic-chickadeedee! saucy note
Out of sound heart and merry throat,
As if to say, 'Good day, good sir!
Fine afternoon, old passenger!
Happy to meet you in these places,
Where January brings few faces.'

This poet, though he lived apart,
Moved by his hospitable heart,
Sped, when I passed his sylvan fort,
To do the honors of his court,
As fit a feathered lord of land;
Flew near, with soft wing grazed my hand,
Hopped on the bough, then, darting low,
Prints his small impress on the snow,
Shows feats of his gymnastic play,
Head downward, clinging to the spray.

Here was this atom in full breath,
Hurling defiance at vast death;
This scrap of valor just for play
Fronts the north-wind in waistcoat grey,
As if to shame my weak behavior;
I greeted loud my little savior,
'You pet! What dost thou here? and what for?
In these woods, thy small Labrador,
At this pinch wee San Salvador!
What fire burns in that little chest
So frolic, stout, and self-possest?

Henceforth, I wear no stripes but thine;
Ashes and jet all hues outshine.
Why are not diamonds black and grey,
To ape thy dare-devil array?
And I affirm, the spacious North
Exists to draw thy virtue forth.
I think no virtue goes with size;
The reason of all cowardice
Is, that men are overgrown,
And, to be valiant, must come down
To the titmouse dimension.'

'Tis good-will makes intelligence,
And I began to catch the sense
Of my bird's song: 'Live out of doors
In the great woods, on prairie floors.
I dine in the sun; when he sinks in the sea,
I too have a hole in a hollow tree;

And I like less when Summer beats
With stifling beams in these retreats,
Than noontide twilights which snow makes
With tempest of the blinding flakes.
For well the soul, if stout within,
Can arm impregnably the skin;
And polar frost my frame defied,
Made of the air that blows outside.'

With glad remembrance of my debt,
I homeward turn; farewell, my pet!
When here again thy pilgrim comes,
He shall bring store of seeds and crumbs.
The Providence that is most large
Takes hearts like thine in special charge,
Helps who for their own need are strong,
And the sky doats on cheerful song.
Henceforth I prize thy wiry chant
O'er all that mass and minster vaunt;
For men mis-hear thy call in Spring,
As 'twould accost some frivolous wing,
Crying out of the hazel copse, Phe-be!
And in winter, Chic-a-dee-dee!
I think old Caesar must have heard
In northern Gaul my dauntless bird,
And, echoed in some frosty wold,
Borrowed thy battle-numbers bold.
And I will write our annals new,
And thank thee for a better clew.
I, who dreamed not when I came here

To find the antidote of fear,
Now hear thee say in Roman key,
Paean! Veni, vidi, vici.

What a wonderful evocation. Emerson entirely revises the Chickadee's song into historic, triumphant bravery, using the single most famous Roman quotation, heroic Caesar crossing the Rubicon. Most all high schoolers learned these words fifty years ago in Massachusetts. I wonder if any except Latin students encounter them now.

Emerson was interim minister of my New Bedford Unitarian Church in 1831, before Frederick Douglass got his first paid job stacking wood for the minister's wife in 1839. This was before Emerson quit the ministry over communion. Our own church member, Mary Rotch, Quaker light within, may have influenced Emerson, for she would leave church during communion, and Emerson expressed gratitude for her. She also left her famous painting, "The Carpenter's Son," to our church. (The later Nineteenth-Century painter, an Emerson namesake, Edward Emerson Simmons.) It was in the Fellowship hall until it was cut out of its frame and lost for a couple years, found in back of our church kitchen refrigerator. Then returned to the Rotch-Jones Duff house, so one must pay to see it. You can still see Ricketson's bust of Emerson in our chancel, when the church is open.

Glossary

amanuensis   A writer of dictation, as used by blind poet John Milton

appoggiatura   Quick grace notes before the melody note

dactyl   Poetic meter, dominant in Latin: long short short

excursus   Detailed explication

glissando   Slurring, as from a slide trombone or violin (without frets)

haruspex   Seer, prophet

homiletic   Sermon, preaching

metempsychosis   Migration of the soul

micturition   The act of urinating

onomastic   Name related

polysemous   More than one meaning

prelapsarian   Before the fall, Adam and Eve

precieux   Seventeenth-Century French poetic style

secant   Cutting through a circle, not a diameter

sforzando   Note played with sudden, strong emphasis

skyey   Resembling sky

sinusoidal   A sine wave, curve with smoothe periodic oscillation

soi-distant   So-called

terrene   Earthen

trochee   Poetic meter, strong-weak, also used in ordering someone

Works Cited

Allen, Francis H. 1993. *Thoreau on Birds: Notes on New England Birds from the Journals of Henry David Thoreau.* Reprint ed. Boston: Beacon Press.

Amato, B., and A. Pardo, eds. *Dizionario Napoletano.* 1997. Milan: Antonio Vallardi.

Armstrong, Edward A. 1973. *A Study of Bird Song.* Reprint ed. New York: Dover.

Ayer, Alfred Jules. 1946. *Language, Truth and Logic.* New York: Dover.

Baird, Theodore. 1946. "Darwin and the Tangled Bank." *American Scholar* 15, no. 4 (Autumn): 477–86.

———. 1966. "Corn Grows in the Night." In *Thoreau's Walden and Civil Disobedience,* edited by Owen Thomas. Norton Critical Edition. New York: W. W. Norton.

Baptista, Luis F. 2000. "What the White-Crowned Sparrow's Song Can Teach Us About Human Language." *Chronicle of Higher Education,* B8.

Battisti, Carlo, and Giovanni Alessio. 1950. *Dizionario Etimologico Italiano.* Firenze: G. Barbèra.

Bean, Michael, and Melanie Rowland. 1997. *The Evolution of National Wildlife Law.* 3rd ed. Westport, Conn.: Praeger.

Beebee, Trevor J. C. 1983. *The Natterjack Toad.* Oxford: Oxford University Press.

Beletsky, Les. 1996. *The Red-Winged Blackbird: Biology of a Strongly Polygynous Songbird.* London: Academic Press.

Bellow, Saul. 2000. *Ravelstein.* New York: Penguin Books.

Bernstein, Leonard. 1966. *The Infinite Variety of Music.* New York: Simon and Schuster.

Bonora, Maurizio. 1994. *Uccelli di campagna.* Verona: Edizioni l'Informatore Agrario.

Bréberuf, Pére Jean de. 1649. *Relations des Jésuites del la Nouvelle-France.* St. Ignace.

Burton, Frederick R. 1969. *American Primitive Music, with Special Attention to the Songs of the Ojibway.* Reprint. Port Washington, N.Y.: Kennikat Press.

Burton, Robert. 1997. *Comment nourrir les oiseaux de l'Améerique du Nord.* Abridged ed. St. Laurent, Quebec: Éditions du Trécarré.

Burtt, Harold. 1967. *The Psychology of Birds.* New York: MacMillan.

Caelius Aurelianus. 1950. "On Chronic Diseases [Methodici Siccensis tardarum Passionum]." In *On Acute Diseases and On Chronic Diseases,* edited and translated by Israel Drabkin. Chicago: University of Chicago Press.

Caldwell, Mark. 2000. "Polly Wanna Ph.D.?" *Discover,* January 1, 69.

Catchpole, C. K., and P. J. B. Slater. 1995, *Bird Song: Biological Themes and Variations.* Cambridge: Cambridge University Press.

Catesby, Mark, FRS. 1731. *The Natural History of Carolina, Florida and the Bahama Islands.* 2 vols. London.

Cheney, Simeon Pease. 1892. *Wood Notes Wild: Notations of Bird Music.* Boston: Lee and Shepard.

Chomsky, Noam. 1966. *Cartesian Linguistics.* New York: Harper and Row.

Cicero, M. Tullius. 1964. *De Senectute; De Amicitia; De Divinatione.* Translated by W. A. Falconer. Loeb Classical Library. Cambridge, Mass.: Harvard University Press.

Cornish, Francis Warre, ed. 1966. *Catullus, Tibullus and Pervigilium*

Veneris. Loeb Classical Library. Cambridge, Mass.: Harvard University Press.

Danforth, Susan, Curator, and William McNeil. 1991. *Encountering the New World 1493–1800.* Providence, R.I.: John Carter Brown Library.

Deacon, Terrence W. 1997. *The Symbolic Species: The Co-Evolution of Language and the Brain.* New York: W. W. Norton.

De Saussure, Ferdinand. 1959. *Course in General Linguistics.* Edited by Bally and Sechehays. Translated by Wade Baskin. New York: Philosophical Library.

De Tovar, Juan. 1585. *Historia de la Benida de los Yndios.* Mexico.

Dickinson, Emily. 1960. *The Complete Poems of Emily Dickinson.* Edited by Thomas H. Johnson. Boston: Little, Brown.

Ehrlich, Paul, David Dobkin, and Darryl Wheye. 1988. *The Birder's Handbook: A Field Guide to the Natural History of North American Birds.* New York: Simon and Schuster.

Elliott, Lang. 1994. *Know Your Bird Sounds,* 2 vols. Cassette audio books. Ithaca, N.Y.: NatureSound Studio.

Emerson, Ralph Waldo. 1885. *Poems of R. W. Emerson.* London: Walter Scott.

Falls, J. Bruce. 1969. "Functions of Territorial Song in the White-Throated Sparrow." In *Bird Vocalizations,* edited by R. A. Hinde. Cambridge: Cambridge University Press.

Fenton, William N. 1978. "Northern Iroquoian Culture Patterns." In *Handbook of North American Indians,* vol. 15, edited by Bruce Trigger. Washington, D.C.: Smithsonian.

Ferguson, Margaret, Mary Jo Salter, and Jon Stallworthy, ed. 1997. *The Norton Anthology of Poetry.* 4th ed. New York: W. W. Norton.

Flood, Renée Sansom. 1995. *The Lost Bird of Wounded Knee: Spirit of the Lakota.* New York: Scribner.

Francis of Assisi. 1995. *Fontes Franciscani.* Edited by Enrico Menesto and Stefano Brufani. Assisi, Italy: Edizioni Porziuncola.

Fuller, Thomas. 1655. *Ornitho-logie, or, the Speech of Birds*. London: Printed for John Stafford.

Green, Julien. 1983. *God's Fool: The Life and Times of Francis of Assisi*. San Francisco: HarperSanFranciso.

Grossinger, Richard. 2020. *Bottoming Out the Universe: Why There Is Something Rather Than Nothing*. Rochester, Vt.: Park Street Press.

Guthrie, Ramon, and George Diller, eds. 1942. *French Literature and Thought Since the Revolution*. New York: Harcourt, Brace and World.

Habig, Marion, ed. 1991. *St. Francis of Assisi, Omnibus of Sources. Writings and Early Biographies*. Quincy, Ill.: Franciscan Press.

Harris, Roy. 1966. *Language, Saussure and Wittgenstein: How to Play Games with Words*. New York: Routledge.

Hartshorne, C. 1956. "The Monotony Threshold in Singing Birds." *Auk* 73: 176–92.

———. 1973. *Born to Sing: An Interpretation and World Survey of Bird Song*. Bloomington: Indiana University Press.

Hay, John, ed. 1996. *The Great House of Birds: Classic Writings About Birds*. San Francisco: Sierra Club Books.

Heizer, Robert, and Albert Elsasser. 1980. *The Natural World of the California Indians*. Berkeley: University of California Press.

Herman, Louis, and Sheila Abichandani. 1999. "Dolphins (*Tursiops truncatus*) Comprehend the Character of the Human Pointing Gesture." *Journal of Comparative Psychology* 113, no. 4 (December): 347.

Herrick, Robert. 1960. *The Poems of Robert Herrick*. London: Oxford University Press.

Hinde, R. A., ed. 1969. *Bird Vocalizations*. Cambridge: Cambridge University Press.

Hirschfelder, Arlene, and Paulette Molin. 1992. *The Encyclopedia of Native American Religions*. New York: Facts on File.

Hoxie, Frederick E., ed. 1996. *Encyclopedia of North American Indians*. Boston: Houghton Mifflin.

Hudson, W. H. 1924. *Afoot in England*. London: J. M. Dent and Sons.

Hutson, H. P. W. 1956. *The Ornithologists' Guide, Especially for Overseas*. London: British Ornithologists' Union.

Kightley, Chris, and Steve Madge. 2002. *Pocket Guide to the Birds of Britain and North-West Europe*. London: Pica Press.

Kircher, Athanasius. 1650. *Musurgia universalis, sive ars magna consoni et dissoni*. Rome: Haenedium Francisci Corbelletti.

Krause, Bernard L. 1998. *Into a Wild Sanctuary: A Life in Music and Natural Sound*. Berkeley, Calif.: Heyday Books.

Kristeva, Julia. 1989. *Language: The Unknown; An Initiation into Linguistics*. Translated by Anne M. Menke. New York: Columbia University Press, 1989.

Kroodsma, Donald E., and Edward H. Miller, eds. 1996. *Ecology and Evolution of Acoustic Communication in Birds*. Ithaca, N.Y.: Cornell University Press.

Lack, David. 1965. *The Life of the Robin*. London: Witherby.

Lawall, Sarah, and Maynard Mack, eds. 1999. *Norton Anthology of World Masterpieces,* vol. 2. 7th ed. New York: W. W. Norton.

Li Bo. 1988. "Waking from Drunkenness on a Spring Day." In *Gems of Classical Poetry in Various English Translations,* edited by Lu Shu-xiang and Xu Yuan-zhong; translated by Arthur Waley. Hong Kong: Joint Publishing.

Locke, John. 1961. *An Essay Concerning Human Understanding*. 2 vols. London: J. M. Dent.

Lorenz, Konrad. 1952. *King Solomon's Ring*. New York: Crowell.

MacDonald, Hugh, Ed. 1966. *The Poems of Andrew Marvell*. Cambridge, Mass.: Harvard University Press.

MacKay, Barry Kent. 2001. *Bird Sounds: How and Why Birds Sing, Call, Chatter, and Screech*. Mechanicsburg, Pa.: Stackpole Books.

Margoliash, Daniel. 2001. "The Song Does Not Remain the Same." *Science* 291, no. 5513 (March 30): 2559–61.

Marler, Peter, Christopher S. Evans, and Marc D. Hauser. 1992. "Animal Signals: Motivational, Referential, or Both?" In *Nonverbal Vocal Communication: Comparative and Developmental Approaches,* edited by Hanus Papousek, Uwe Jurgens, and Mechthild Papousek. Cambridge, UK: Cambridge University Press.

Mastro, Jim G. 1999. "Signal to Noise." *Sciences* 39, no. 6 (November): 32.

Mathews, F. Schuyler. 1910. *Field Book of Wild Birds and Their Music.* New York: Putnam.

Merriam, C. Hart. 1882. "List of Birds Ascertained to Occur within Ten Miles from Point de Monts, Province of Quebec, Canada; Based Chiefly upon the Notes of Napoleon A. Comeau." "General Notes." *Bulletin of the Nuttall Ornithological Club* 7, no. 4 (October): 233–42, 120–21.

Messiaen, Olivier. 1994. *Olivier Messiaen: Music and Color, Conversations with Claude Samuel.* Translated by E. T. Glasow. Portland, Ore.: Amadeus Press.

———. *The Technique of My Musical Language.* Translated by John Satterfield. Paris: A Leduc, 1956–61.

Messiaen, Olivier, and Claude Samuel. 1994. *Olivier Messiaen: Music and Color: Conversations with Claude Samuel.* Translated by E. Thomas Glasow. Portland, Ore.: Amadeus Press.

Metz, R. C. 1990. *Langenscheidt's Standard Italian Dictionary.* New York: Bantam Books, 1990.

Michelet, Jules. 1876. *L'Oiseau.* 2nd ed. Paris: Hachette.

Milius, Susan. 2000. "When Ants Squeak." *Science News* 157, no. 6 (February): 92.

Miller, Olive Thorne. 1885. *Bird-Ways.* Boston: Houghton Mifflin, 1885.

Morris, D. 1989. "A Semiotic Investigation of Messiaen's '*Abime des oiseaux.*'" *Analles Musique,* viii: 125–58.

Neruda, Pablo. 1985. *Art of Birds*. Translated by Jack Schmitt. Austin: University of Texas Press.

———. 1991. *Canto General*. Translated by Jack Schmitt. Berkeley: University of California Press.

Nice, Margaret. 1964. *Studies in the Life History of the Song Sparrow*. 2 vols. New York: Dover, 1964.

Nottebohm, Fernando. 1981. "Brain Pathways for Vocal Learning in Birds: A Review of the First Ten Years." In *Progress in Psychobiology and Physiological Psychology*, vol. 9, edited by M. S. Sprague and A. N. E. Epstein, 85–124. New York: Academic Press.

Olschki, Leonardo. 1937. *Storia Letteraria delle Scoperte Geografiche*. Florence, Italy: Olschki.

Paliotti, Vittorio. 1992. *Storia della canzone napoletana*. Rome: Newton Compton Editori.

Pasquier, Roger F. 1977. *Watching Birds: An Introduction to Ornithology*. Boston: Houghton Mifflin.

Pepperberg, Irene M., and Sarah E. Wilcox. 2000. "Evidence for a Form of Mutual Exclusivity during Label Acquisition by Grey Parrots (*Psittacus erithacus*)." *Journal of Comparative Psychology* 114, no. 3 (September): 219–31.

Peterson, Roger Tory. 1947. *A Field Guide to the Birds: Eastern Land and Water Birds*. Boston: Houghton Mifflin.

———. 2010. *Peterson Field Guide to Birds of Eastern and Central North America*. 6th ed. New York: Houghton Mifflin Harcourt.

———. 1999. *Le Guide des Oiseaux du Québec et de l'Est de L'Amérique du Nord*. Traduction de Blain, Cyr, Normand, et Gosselin. Boucherville, Québec: Broquet.

Peterson, Roger Tory, Guy Mountfort, and P. A. D. Hollom. 2001. *Birds of Britain and Europe*. 5th ed. New York: Houghton Mifflin Harcourt.

Pinker, Steven. 2000. *The Language Instinct: How the Mind Creates Language.* New York: Harper Perennial.

Powers, Alan W. 2013. "Giordano Bruno and the Search for Habitable Worlds." Talk at Harvard-Smithsonian Center for Astrophysics. YouTube video, 23:27.

———. 2010. "Giordano Bruno and the Seicento Moon-mappers Galileo and Riccioli." In *The Worlds of Giordano Bruno.* Birmingham, UK: Cortex Design.

———. 1987. "On Language: Head over Googol." *New York Times Magazine,* section 6, 10.

———. 1994. *Westport Soundings.* Westport, Mass.: A. Powers.

Powers, William K. 1986. *Sacred Language: The Nature of Supernatural Discourse in Lakota.* Norman: Oklahoma University Press.

———. 1970. "Songs of the Red Man." *Ethnomusicology* 14, no. 2: 358–69.

Read, Gardner. 1990. *20th-century Microtonal Notation.* Westport, Conn.: Greenwood Press,.

Rosinsky, Natalie M. 2001. "'Going Ape' over Language." In *Odyssey: Adventures in Science,* vol. 10, no. 7. Peterborough, N.H.: Cobblestone.

Saunders, Aretas A. 1935. *A Guide to Bird Songs.* New York: D. Appleton-Century.

Saussure, Ferdinand de. 1959. *Course in General Linguistics.* Edited by Charles Bally and Albert Sechehaye. Translated by Wade Baskin. New York: Philosophical Library.

Schleidt, W., et al. 1960. "Störung des Mutter-Kind Beziehung durch Verhöverlust." *Behavior* 16: 254–60.

Shakespeare, William. 1992. *William Shakespeare: The Complete Works.* Edited by Stanley Wells and Gary Taylor. Oxford, UK: Oxford University Press.

Shepard, Paul. 1999. *Encounters with Nature.* Edited by Florence R. Shepard. Washington, D.C.: Island Press.

Simmers, R. W. 1975. "Variations in the Vocalizations of Male Red-winged Blackbirds (*Agelaius phoeniceus*)." Ph.D. thesis, Cornell University, Ithaca, New York.

Skutch, Alexander F. 1996. *The Minds of Birds*. College Station: Texas A&M University Press.

Smith, Anne M., ed. 1993. *Shoshone Tales*. Salt Lake City: University of Utah Press.

Smith, Douglas G., and Fiona A. Reid. 1979. "Roles of the Song Repertoire in Red-winged Blackbirds." *Behavioral Ecology and Sociobiology* 5: 279–90.

Specter, Michael. 2001. "Rethinking the Brain: How the Songs of Canaries Upset a Fundamental Principle of Science." *New Yorker,* July 23: 42–53.

Sprague, James M., and Alan N. Epstein, eds. 1980. *Progress in the Psychobiology and Physiological Psychology,* vol. 9. New York: Academic Press.

Sturtevant, William, ed. 1978. *Handbook of North American Indians,* vol. 13. Washington, D.C.: Smithsonian Institution.

Swadesh, Morris. 2006. *The Origin and Diversification of Language*. London: Routledge.

Swift, Jonathan. 1967. *Gulliver's Travels and Other Writings*. Edited by Miriam Starkman. New York: Bantam Books, 1967.

Taylor, Hollis. 2017. *Is Birdsong Music? Outback Encounters with an Australian Songbird*. Bloomington: Indiana University Press.

Taylor, Kip. 1988. *Loon*. Saranac Lake, N.Y.: Kip Taylor.

Taylor, Marianne. 2018. *RSPB British Birds of Prey*. Bloomsbury: Wildlife.

Thielcke, Gerhard A. 1976. *Bird Sounds*. Translated by John Drury. Ann Arbor: University of Michigan Press.

Thorpe, W. H. 1961. *Bird-Song*. Cambridge, UK: Cambridge University Press.

Thundup, Tulku. 1998. *The Healing Power of Mind.* Rochester, Vt.: Inner Traditions.

Waldbauer, Gilbert. 1998. *The Birder's Bug Book.* Cambridge, Mass.: Harvard University Press.

Warren-Chadd, Rachel, and Marianne Taylor. 2016. *Birds: Myth, Lore and Legend.* London: Bloomsbury.

———. "Delving into Cultural Myths, Tales and Beliefs about Birds." Interview with Rachel Warren-Chad and Marianne Taylor, National Geographic Newsroom. National Geographic Blog.

Weed, Clarence, and Ned Dearborn. 1903. *Birds in Their Relation to Man.* Philadelphia and London: J. B. Lippincott.

Williams, Roger. 1971. *A Key into the Language of America.* Reprint. Ann Arbor: Gryphon Books, 1971. First published 1643.

Yasukawa, Ken. 1979. "Territory Establishment in Red-Winged Blackbirds: Importance of Aggressive Behavior." *Condor* 81: 258–64.

Young, Jon. 1999. *Advanced Bird Language.* Audiocassette. OWLink Media.

Gull

Acknowledgments

*I*n addition to those listed in my notes, I have received a wide variety of help. I am indebted to librarians at the British Library and at Brown University—particularly those at the Science Library and Susan Danforth at the John Carter Brown Library—as well as to Dr. Gabriela Adler at my own Bristol Community College Library. This book could not have been written without the inspiring conversation of Richard C. Wheeler. I am indebted to music instruction in computerized notation from Prof. Andrew McWain, to a crucial music loan from Judy Conrad, and to operatic discussions with Richard Clark. I have prospered under the philosophical tutelage of Dr. Bill Kaufmann. My conversation has improved from long association with the raconteur Michael Miller.

One could not have settled among more generous and expert neighbors than Rick and Marge McNally, Jennifer McIntire, Marie Hadfield, Harriet and Bill Barker, and Selena and Jim Howard. Birders such as Philip Sheehan and Kevin and Deborah Lawton have fostered my interests. The writers Jamie O'Neill and Tim Coutis have inspired me, as have the scientists Drs. Jon and Liz Wolpaw, and Dr. Dave Warr. My colleagues Art Lothrop, Joe Murphy, and John Majkut lighten our daily teaching load. Former students like Mark Bisson, Mark Folco, Omayra Fontanez, and Mark Sardinha leave me grateful. Finally, Isaac and Laura Segal, Rev. David M. Powers, Marie Ged and Dan Mallett, Dr. C. "Ben" Lindsay, and Dr. Jean D'Amato Thomas have been unwavering supporters.

It is customary to record last one's greatest debt, to the creature (not a bird in this case) who illustrated and illuminated this manuscript and my life.

Index of Birds

General Index

About the Author

Alan Powers graduated from Amherst College in 1966 and earned his Ph.D. in English from the University of Minnesota in 1974. Since that year he has been on the faculty at Bristol Community College, where he was chairman of the English Department from 1988–1992. A scholar of Renaissance Italian, English, and American literature, he has published and lectured widely on Shakespeare, Giordano Bruno, Seventeenth-Century and Romantic poets and poetry, and Emily Dickinson and other American writers. Interested as much in local history as language, he researched and published on early American New England street games.

He has held many fellowships and prizes for research and his own poetry, and in 2001 made a guest appearance in James Wolpaw's documentary film on Emily Dickinson.

Powers lives in Westport, Massachusetts, with his wife, the artist Susan Mohl Powers.